SALES & MARKETING CHECKLISTS FOR PROFIT-DRIVEN HOME BUILDERS

3rd Edition

Sales and Marketing Checklists

for Profit-Driven Home Builders, Third Edition

Jan Mitchell

BuilderBooks, a Service of the National Association of Home Builders

Elizabeth M. Rich	Director, Book Publishing
Natalie C. Holmes	Book Editor
Color of Paradise	Cover Design
Electronic Quill Publishing Services	Composition
Gasch Printing	Printing

Gerald M. Howard	NAHB Chief Executive Officer
Mark Pursell	NAHB Senior Vice President, Expositions, Marketing & Sales
Lakisha Woods, CAE	NAHB Vice President, Publishing & Affinity Programs

Produced in the United States of America

18 17 16 15 14 1 2 3 4 5

ISBN: 978-0-86718-730-4
eISBN: 978-0-86718-728-1

Cataloging-in-Publication Information

Library of Congress CIP information available on request.

For further information, please contact:

National Association of Home Builders
1201 15th Street, NW
Washington, DC 20005-2800
800-223-2665
BuilderBooks.com

CONTENTS

IN MEMORIAM

Jan Mitchell, the author of the book you're reading, passed away after a two year battle with GIST, an uncommon type of gastric cancer, in September 2013. She had put the finishing touches on this book shortly before her death. She'd been a friend and colleague of mine since I hired her to work as the Assistant Director of the NAHB's National Sales & Marketing Council (NSMC) in 1987. The first time I met her, I was struck by her easy-going demeanor, her Texas twang (she was a proud Aggie!), her personal style and her professionalism, all wrapped into one terrific package. I knew immediately that she would be a fantastic member of the NSMC team, but had no idea that we would be friends for life. Her friendship ended up being one of my life's most important ones and I am eternally grateful to have had it.

Jan left NAHB in 1990, but continued to work for NAHB/NSMC as the Senior Editor of *Sales + Marketing Ideas* and in the building industry until her death. Since leaving NAHB, she wrote/edited four books for NAHB BuilderBooks and its predecessor, Home Builder Press: the best-selling *Sales and Marketing Checklists for Profit Driven Home Builders* (1st edition in 1997, 2nd in 2003 and 3rd in 2014) and *The Best of Sales & Marketing Ideas* (1999). Additionally, she was a freelance writer for numerous building industry publications, including *BUILDER*, *Custom Builder*, *Professional Builder* and, of course, the publication near and dear to her heart, *Sales + Marketing Ideas*. She absolutely loved to write the annual article about The Nationals' (National Sales and Marketing Awards) merchandising winners, which married her passion for writing and editing with her passion for decorating. She definitely had an eye for style!

Beyond the professional, Jan had a warmth about her that very few people have. She was genuine, had a great sense of humor and incredible integrity. When she asked how you were, it wasn't a pleasantry; she really wanted to know—in detail! She was deeply interested in the world around her—she was a passionate world traveler—and particularly in her local community, where she volunteered in a variety of leadership positions, but especially loved working in the local schools.

On behalf of the whole team at NAHB, we are grateful for Jan's dedication and service to the industry, and offer personal thanks for sharing her talents and passion with us and all she did to make the industry a better place.

If Jan were writing this dedication, I have no doubt she would have ended it by thanking her family: her mother, Olivia, her father, Jerry (deceased) and her two siblings, Jay and Joanne—they gave her great joy—and, especially, her adoring (and adored) husband, Ken, and her wonderful sons, Austin and Hayden. While for the last 20+ years, Jan and I were far apart in miles, we often burned the airwaves at length by phone and the age of Facebook allowed me to continue to peek into her life and watch as her boys turned into interesting and kind young men. She had many qualities that I admired, but I particularly admired her relationship with Ken and her boys. She loved them well and was well-loved in return. As our mutual friend, Melissa Bailey, said in a letter to Jan recently, "yours is a life well-lived." Indeed, I can't say it any better myself.

—Meg Meyer, Vice President, Marketing, NAHB

ACKNOWLEDGMENTS

I'd like to extend a warm thank you to the following friends and professionals whose help and contributions were invaluable in the third edition of this book.

S. Robert August, MIRM, North Star Synergies; Doris Pearlman, MIRM, Possibilities for Design; David Miles, MIRM, Miles BrandDNA; Meredith Oliver, Creating WOW! Communications; Melinda Brody, Melinda Brody & Associates; Carol Ruiz, NewGround PR &Marketing, Mary DeWalt, Mary DeWalt Design Group; Angela M. Harris, TRIO Environments; Cassandra Grauer, Jeff Shore Consulting; and Joan Marcus-Colvin, The New Homes Company. A special thanks to Joe McGaw and Sonora Munks of NAHB's NSMC and to Lisa Parrish, Partner Peter Mayer Productions and Administrator of NAHB's "The Nationals" Awards competition.

The following contributed to all previous versions of this book: Angi Ma Wong, Intercultural & Feng Consulting and Corporate Training; Lita Dirks, MIRM, Lita Dirks & Associates; Jay Goldberg, MIRM, D.R. Horton Homes; Carol Ruiz, MIRM, NewGround PR; Tim Kane, MBK Homes, Inc.; Richard Elkman, MIRM, Group Two Advertising; Rita Ortloff, Ideal Homes; S. Robert August, MIRM, North Star Synergies, Roger Fiehn, MIRM, Roger Fiehn & Associates; Sharon Dyer, 5th Gear Advertising; Tom Specht, BonoTom Studio

The following people reviewed the outline and/or part or all of the manuscript and provided suggestions for improving this book: Linda Berger, Chairman, NSMC Builder-Broker Committee, and Vice President of Marketing, Woodcraft Homes, Fort Collins, Colorado; Carol Ann Cardella, MIRM, President, Cardella and Associates, Cincinnati, Ohio; Rhonda Daniels, Director, NAHB Office of Regulatory Control; Lolita (Lita) Dirks, MIRM, President, Lita Dirks and Company, Englewood, Colorado; Roger G. Fiehn, MIRM, President, Roger Fiehn and Associates, Houston, Texas; John Gomall, Editor and Publisher, Housing Executive Report, Lewes, Delaware; David S. Jaffe, NAHB Staff Counsel; Meg Meyer, Executive Director, NAHB National Sales and Marketing Council; Suzanne Meteyer, Real Estate Agent, Manning Realty, Waldorf, Maryland; Yvonne Rawson, MIRM, Vice President, Pueblo Builders, Inc., Las Cruces, New Mexico; E. Lee Reid, President, E. Lee Reid and Company, Apollo Beach, Florida; Phillip Lee Russell, President, Energy Smart Corporation, Pensacola, Florida; Page Snyder, Council and Committees Activities Coordinator, Home Builders Association of Richmond, Virginia; Diane and Duane Willenbring, Secretary-Treasurer and President respectively, Willenbring Construction, Inc.; St. Cloud, Minnesota; B.J. Young, MIRM, MRA, CRB, President, B.J. Young Incorporated, Winter Park, Florida.

Jan Mitchell
July 2013

INTRODUCTION

Since the first edition of this book was published in 1997, the marketing of new homes has drastically changed. Early adopters of the Internet were just dipping their toes in the marketing waters, and customer relationship management (CRM) was largely accomplished via index cards housed in a physical tickler file. A "smart" phone was a chunky mobile device that would save the last call entered!

The second edition of *Sales and Marketing Checklists for Profit-Driven Builders*, updated in 2003, included a new chapter on utilizing technology and counseled home builders that an online presence would soon be just as important as having a phone number. That was just one year before a kid named Mark Zuckerberg would start up a social network at Harvard University called Facebook.

Customers were beginning to search for homes on the Web before they ever set foot into a sales office or model home. At that time, we cited a 2001 study that showed 42% of home buyers used the Internet as a shopping source. By 2013, that number had rocketed to 90%.[1]

Fast forward to 2014, and it is clear that the material in this edition, or in any book of this type, is likely to need updating as fast as we can write it. The world we live in, do business in, and sell homes in is simply changing too fast to keep up. That being said, home builders can stay as educated, informed, and up to date on current business practices and the latest innovations as possible, and this third edition of *Checklists* will help you achieve that goal.

Another continuing challenge for home builders is the changing face of their buyers. The U.S. Census Bureau showed one of the biggest changes between 2000 and 2010 was in the increase in households headed by women without husbands—up by 18% in the decade.[2] Blended families are ever more prevalent, and a greater need for retirement housing is seen as baby boomers reach their golden years. And the wave of immigration continues, making an understanding of multicultural markets more important than ever.

Many home builders were finally breathing a sigh of relief as we emerged from the turbulent economy of 2008 to 2011, but their newest struggle may very well be filling pent-up demand and rehiring labor that may have switched to other industries. And for almost all, tighter lending standards, more discerning customers, and threats to eliminate the mortgage interest deduction are challenges to overcome.

Whatever the challenge in your market, by following the guidance in these checklists, you can make your company more visible, increase your market share, and establish your credibility as a builder poised to do business with today's and tomorrow's prospects.

How to Use This Book

The checklists and forms in this book can guide you through the steps in developing and carrying out your marketing plan. You can use these lists, forms, surveys, etc., as a guide to create files on your own company letterhead or otherwise customize them to fit your particular needs.

You may discover that you already practice some of what this book recommends so you can check those tasks off the lists. Other ideas may be new. Some of the suggestions may not be feasible for you because of local market conditions and customs. But consider everything; you may discover new and creative possibilities for meeting your marketing challenges.

Sales and Marketing Checklists for Profit-Driven Home Builders specifically targets small- to medium-volume builders, but any builder could use most of the ideas.

The book also includes

- Marketing terms italicized in the text and defined in a glossary at the back of the book
- "Tips from a Pro" with ideas and examples from some top sales and marketing professionals in the industry
- "Budget-Cutting Tips" throughout
- "Tech Tips"
- "Read More About it" recommendations

Because marketing efforts and sales are interrelated, to maximize the benefits of this book you should review it in its entirety before beginning your marketing plan. Good luck and happy marketing!

FIGURES

BEGIN WITH MARKET RESEARCH

1

Before breaking ground on a new home, a builder who expects to be successful considers the location, product, and price range that will attract specific types of buyers. If a builder misses the targeted buyer or miscalculates the market because of inaccurate information, a home may sit unsold for months longer than anticipated, forcing the builder to dig deeper and deeper into cash reserves to meet monthly interest payments. Whether you build five homes a year or 50, a successful building company begins with a marketing plan based on solid research.

Although it sounds complex, don't let the term *market research* intimidate you. It simply entails gathering some basic information about your marketplace, your buyers, their lifestyles and preferences, and your competition before you start a building project. You will use this data to

- develop a buyer profile;
- determine the type of house to build;
- establish a price range;
- choose a location;
- create a marketing strategy to attract buyers; and
- generate referrals and sales until your project sells out or the home is sold.

Gathering the information to inform your decisions is neither costly nor complicated and will save you money, headaches, and heartache in the long run.

Research Your Buyers and Their Location

You cannot build for a market unless you understand it. Following are suggestions about where you can gather useful data to inform your project planning.

Contact the U.S. Census Bureau

The U.S. Census Bureau compiles a wealth of information about the people who live in *metropolitan statistical areas.* You can use the American Fact Finder at http://factfinder.census.gov to get population, housing, economics, and geographic data based on the 2010 (or most recent) Census.

On its website, you can obtain a variety of different reports, such as a population profile for any street or block, business patterns for any zip code, or a county profile. For more general information on any city or county, go to Quick Facts at quickfacts.census.gov/qfd. For further help or other requests, call 800-923-8282.

Tech Tip

Download the free mobile app from the Census Bureau to get quick facts right on your smart phone. Go to census.gov

For ongoing free updates from the census, follow the U.S. Census Bureau on Twitter at twitter.com/uscensusbureau. Using census data, you should be able to answer the following questions:

- What age, category, or categories represent the largest numbers of residents in the researched area?
- What is their average income?
- What is the typical number of household members?
- What is the typical family or marital status of the people in these households?
- How many children does the typical household include, and what ages?
- How many years of education do the adults typically have?
- What percentage of households own their homes and what percentage currently rent their homes?
- What are the largest sources of local employment?

Contact Your Local Chamber of Commerce

Ask for its community profile or relocation package. These profiles or packages provide information about the local business climate and are usually free or available for a nominal charge. Of course, the type of information available varies from one chamber to another, but all usually will include the answers to the following questions:

1. What percentage of annual population growth did the area experience over the last year, and what is projected for the coming year?
2. What local businesses may be transferring people into the area?
3. What new corporations may be planning to relocate to your community? How many people will they add to the community? What types of workers and at what relative income levels will be moving in, and where will the new company build its facilities?
4. In what types of jobs does the majority of the population work (construction, high-tech, manufacturing, financial, insurance, real estate, professional services, government, other)?
5. What is the current unemployment rate (or most recent data available)?
6. How many new jobs did that percentage represent during the last year (or latest year data was available)?
7. How many new homebuilding permits has the local jurisdiction issued during the last year?
8. Calculate the employment to permit ratio (E:P ratio) for your area as follows: Divide the number of new jobs created during the last 12 months (see question 6) by the number of new homebuilding permits pulled during the same time period (see question 7). Nationally, an E-P Ratio of 1.8 to 1 indicates a comfortable level of absorption for new home builders.
9. Who are the area's major employers? Most chambers of commerce generally list the top 10, 20, or 100 employers in their cities or metropolitan areas according to how many people these enterprises employ. The primary types of workers they employ and the part of town in which those employers are located indicates of the type of housing that will be needed in your area. Record the top 10 employers for your community below:

	Company	Types of Employees	Location
1.	____________	____________	____________
2.	____________	____________	____________
3.	____________	____________	____________
4.	____________	____________	____________

5.			
6.			
7.			
8.			
9.			
10.			

Conduct Customer and Prospect Surveys

Two ways to accumulate valuable information about your prospects are visitor registration cards and after-move-in surveys.

CUSTOMER RELATIONSHIP MANAGEMENT (CRM)

A rigorous registration policy will help capture information on every visitor to your model, speculative home, or sales office. If you employ on-site sales representatives, train them to gather information on each customer and to ask the questions conversationally. Assure visitors that the information will not be sold to a third party and will be used strictly to ensure that you are building the correct product for the right market.

Once you capture the information, enter each contact into a Customer Relationship Management software program. A good one will provide you with timelines for following up with customers and templates for letters and email. Form 1.1 shows what information you should be capturing and entering on each customer and prospect.

Periodically review the information in your prospect registration data and a picture of your prospects will begin to emerge.

AFTER MOVE-IN SURVEYS

Send postcard or email surveys to all customers 30 days after move-in, 90 days after move-in, and again at the 1 year mark. Over time, these surveys will yield valuable information about your customers' satisfaction levels and about the benefits, features, and floor plans that are most successful. Sample questions include "How would you rate the performance of our customer service and warranty program," "Would you purchase one of our homes again," and "Would you feel comfortable referring others to us?" (See Chapter 13, Form 13.2, Sample Customer Service Survey, for a more complete list of questions to include).

Monitor Local and Regional Publications

Constantly monitor the following publications, whether print or online, for trends and opportunities, such as companies transferring employees into an area:

- Business and real estate sections of local newspaper(s) and local online news sites
- Local home builders association magazine or newsletter
- Local business magazines
- Local chamber of commerce publication

Form 1.1 Prospect Registration

Name(s) ______________________ Address ______________________

City ______________ State _____ Zip ________ Phone ______________

Email address ____

1. Where are you living now? ______________________________
2. Do you own or rent your present home? ○ Own ○ Rent
3. How soon do you plan to relocate?
   - ○ Immediately
   - ○ Within 6 months
   - ○ Within 1 year
   - ○ Longer than 1 year
4. How many people are in your household? ______________________
5. In what price range are you looking for a home? Note: Adjust price ranges as needed for your market area.

   ○ Under $100,000 ○ $250,001–$300,000
   ○ $100,000–$150,000 ○ $150,001–$200,000
   ○ $200,001–$250,000 ○ $300,001–$400,000
   ○ $400,001–$500,000 ○ Over $500,000
6. What is your occupation? ______________________________
7. How did you learn about our homes?
   - ○ Driving by
   - ○ Billboards or directional signage
   - ○ Newspaper ad. Which paper? ______________________
   - ○ Magazine ad. Which magazine? ______________________
   - ○ Radio ad. Which program and station? ______________________
   - ○ Television ad or cable listing. Which channel? ______________________
   - ○ A friend's referral. Who? ______________________
   - ○ A Realtor's® referral. Who? ______________________
   - ○ Our Website. What led you to our site? ______________________
   - ○ Homes search portal on Internet. Which One? ______________________
   - ○ A mailing to your home
   - ○ Other. Specify ______________________

Form 1.1 Prospect Registration (*continued*)

For builders without speculative or model homes:

8. What interests you most about our homes after meeting with the builder or sales representative? ______

____

____

9. What, if anything, would you like to change about the floor plans, designs, or renderings you saw? ____

____

____

For builders with speculative or model homes:

8. What did you like most about our homes after touring our model/speculative home? ________________

____

____

9. What, if anything, would you like to change about the model/speculative home you saw ____________

____

____

For all builders:

Salesperson's Use Only:

Contact Priority: A B C D

Conduct Neighborhood Audits

In perhaps one of the more enjoyable tasks in marketing research, especially for those who enjoy getting "out in the field," drive and walk around in the new home communities in your area. Study the demographics and design trends of each neighborhood in the area(s) where you build or are planning to build. This examination will help you determine what kinds of buyers live in your area. These people may become your next move-up customers. Note the following:

- Typical number of cars owned and type
- Evidence of children? (For example, do you see play equipment in yards?)
- Stay-at-home parents or dual-income households? (Can you see a parent at home or do homes appear to be empty during the day?
- What typical age(s) do owners appear to be?
- Designs noted
- Number of stories
- Number of garage stalls, tandem garage: yes or no
- Approximate square feet
- Approximate lot size
- Primary architectural style: Colonial, Contemporary, Midcentury Modern, Craftsman, French Country, Ranch, Southwest, Spanish, Split-level, Tudor, or other

After conducting many of the steps described above, a picture of your buyer profile should begin to emerge.

Composite Builder: Ken Smith

Introducing Ken Smith of KS Builders, a fictional small firm building three to seven homes per year on a speculative basis. Throughout the book we will follow Ken throughout his marketing plan. He began by trying to determine where he should concentrate his efforts in the next year or two. He subscribed to several local online news sources and consulted his local chamber of commerce data. These sources revealed that a computer software manufacturing and service facility in the southwest part of his town was undergoing a major expansion and would transfer in 5,000 new middle-management employees within the next year.

Then Ken consulted census data for two zip codes in the southwest section. It revealed that residents were predominantly college-educated, married-couple households with an average of two children.

Ken determined that the southwest part of town is probably a viable location in which to purchase home sites for building, and that he should build in a move-up price range to target increasingly successful business executives with growing families. Still he wanted a little more information about their lifestyles and preferences and the competition that is currently building in the area.

Understanding Your Buyers' Lifestyles

Now that you know roughly who and where your buyers are, the next step is to get a detailed profile of them and learn as much as possible about their lifestyles, just as Ken Smith is planning to do in the example. That way, you will be able to build homes that match your targeted buyers' dreams and desires. A *focus group* is one of the best ways to obtain this information. It is a discussion group of carefully-chosen participants gathered together to provide *qualitative data* from their responses and opinions on a specific topic. No matter what your annual sales volume is, focus groups are not out of your reach. You can make them complicated or simple, expensive or reasonable, depending on a number of factors. See Checklist 1.1 to learn what is involved in holding a focus group.

Checklist 1.1 Focus Groups

___ Participants: 10–12. Invite three or four more than you will need to allow for no-shows and last-minute emergencies.

___ Facilitator. Choose a group discussion leader. This person must be able to ask the questions in such a way that will produce detailed, objective responses and will keep the discussion on track and moving forward. You can do this yourself if you feel comfortable in the role and believe you can be totally objective or you may want to hire a professional market researcher. (Consult the NAHB Directory of Professionals with Home Building Designations at http://tinyurl.com/knb9edq.)

___ Location. In some areas, you can rent space in a market research facility, with one-way mirrors and recording equipment. For tighter budgets, your sales office or a conference room in a local hotel, church, or other meeting hall will also do.

___ Refreshments. Provide coffee and bagels or a casual lunch.

___ Audio or video recording. While video recording can be valuable, it may inhibit participants' responses. (Digital audio recorders are easy to use, pick up conversational sound well, and are available for as little as $25 at an electronics store.) You may want to use two recorders so you will have a backup.

___ Offer a small incentive to participate, such as cash or a gift certificate. For real estate agents, coupons for car detailing are enticing.

Your Focus Group Budget

Budgets for focus groups vary widely depending upon a number of factors, such as whether you do the moderating yourself or hire a professional, conduct them at a research facility or at your own site. Table 1.1 compares possible budgets at the low and high end.

Focus Groups with Real Estate Agents

In many areas of the country, the real estate community has a great deal of knowledge about the buyers who are your potential customers. Real estate agents may be particularly insightful about the relocation market.

Foster working relationships with the real estate community in your area. Usually, if a builder has a reputation for fully cooperating with brokers on sales, and agents perceive the builder as a good source of sales for their clients, they will be happy to contribute a few hours of their time to participate in a face-to-face group discussion in spite of their busy schedules. (For more information on builder-broker cooperation, see Chapter 4 Staff for Sales Success.)

Table 1.1 Sample Focus Group Budget

Facility	As low as $0 if you use your own conference room or other free public facility. As high as $700 to $5,000 per group for a formal research facility.
Moderator	As low as $0 if the builder, sales manager, or marketing director moderates the group, or try local organizations, such as Toastmasters International, whose members may be willing to volunteer. As high as $2,500 plus $500 expenses for a professional moderator.
Lunch/Coffee/Refreshments	As low as $50 for coffee and bagels. As high as $300 for full lunch. (Note: Hotels may furnish a conference room for free if you purchase food and beverages from them.)
Incentives	Plan to spend $25 to $100 per participant depending on location. You could use a cash incentive, gift certificate to a local restaurant, or giveaways with your builder logo, such as t-shirts or mugs.
Recording Equipment	As low as $100 if you use your own equipment. As high as $1000 if you rent or order through a hotel or other conference facility.
Report Transcription	About $300 to $500.
Total	From $700 to $11,800.

Conduct Focus Groups with Previous Customers

Assemble a group of buyers who have purchased a home from you within the last year. To ensure their candid responses, you may want to ask a third party or a professional moderator (search the NAHB Directory of Professionals with Home Building Designations at http://tinyurl.com/knb9edq to find one) to ask the questions of this group found in Form 1.2.

Focus Groups With Prospects

Invite prospects who registered their information with you, as well as prospects from area real estate agents. (Agents are more likely to provide you with names if you have an ongoing, professional relationship with the broker community.) Include some previous prospects who ended up buying from another builder so you can find out why. Offer participants incentives, such as $50 to $100 gift card to a popular area restaurant to compensate them for their time. Form 1.3 lists questions to ask this group.

Budget-Cutting Tip

The late Briggs Napier, a new homes marketing consultant, recommended this shortcut for small-volume builders with limited marketing budgets: To get information on your prospects' lifestyles and preferences, conduct a series of what he called *troikas,* small informal discussion groups of about three participants. Put the results of three or four troikas together over a period of time, and you have information similar to that of a larger focus group.

Form 1.2 Focus Group Questions for Previous Customers

1. What have you enjoyed most about your new home? ______________________________

____

____

2. What things would you change about your new home? ______________________________

____

____

For questions 3 and 4, indicate how satisfied you are on the following scale:

Very Satisfied	Satisfied		Dissatisfied	Very Dissatisfied
1	2	3	4	5

3. Overall, how satisfied have you been with your new home?

Very Satisfied	Satisfied		Dissatisfied	Very Dissatisfied
1	2	3	4	5

4. Overall, how satisfied have you been with this builder's customer service and warranty program?

Very Satisfied	Satisfied		Dissatisfied	Very Dissatisfied
1	2	3	4	5

5. Have any problems that have occurred been handled courteously and in a timely manner?
○ Yes ○ No

6. Have the problems been fixed to your satisfaction? ○ Yes ○ No

If not, please explain ______________________________

____

____

____

7. Would you purchase a home from this builder again? ○ Yes ○ No

8. Have you ever referred others to this builder? ○ Yes ○ No

9. Would you feel comfortable referring others to this builder? ○ Yes ○ No

If not, please explain ______________________________

____

____

____

Form 1.3 Focus Group Questions for Prospects

1. How many new homes or new home communities have you looked at in the last 6 months? ________
2. Which area builders are most memorable to you? Why? ________

   ____
3. Have you purchased a new home in the last year? ○ Yes ○ No

   If so, from what builder? ________

   A. For those who responded "yes" to Question 3, what made you choose that builder? ________

   ____

   ____

   B. What could we have done differently to get you to buy from us instead? ________

   ____

   ____
4. After reviewing our floor plans, which ones appeal to you most and why? ________

   ____

   ____
5. Which rooms do you use the most in your home? ________

   ____
6. What are your design and style preferences? ________

   ____
7. Tell us about your commuting patterns: Where do you live? ________

   Where do you work? Shop? Where else do you travel frequently? ________

   ____
8. What are your favorite leisure activities? ________
9. If you had a choice of enjoying extra square footage in the master bedroom or the living areas, which would you prefer? ________
10. Circle the area amenities most important to you: schools, work, churches, shopping, restaurants, parks, airport, other
11. Ask other questions as pertinent to your market. ________

    ____

    ____

Compile Your Results

The example below shows how Ken Smith interpreted and applied the information he gained through focus groups.

Composite Builder: Prospects' Lifestyles and Preferences

Ken Smith conducted a focus group consisting of 11 real estate agents with whom he had recently cooperated on sales or who had listed his homes in the past. He had invited 14, knowing a few would not be able to make it. He conducted the group in the dining room of his partially-furnished speculative home during an evening when no prospects would be likely to visit. He purchased sandwiches and soft drinks from a local caterer and asked the questions himself.

The agents gave him some useful information about their move-up clients' lifestyles and preferences. They said that those who were searching for new homes overwhelmingly wanted large kitchens open to the family room with eat-in islands, lots of storage space, and anything else that would help save time in their busy, active lifestyles. Other time savers they requested were laundry rooms with a sink and generous folding counters, and they wanted their kitchens located right off the garage for ease in unloading groceries. Desires in owners' suites included anything that could give couples busy with careers and parenting some romance and privacy, such as soaking tubs, dual-head showers, and fireplaces. Popular requests for children's bedrooms included large closets, preferably walk-ins. Three-car garages and computer niches were high on their lists as well.

Ken found out that two builders who these agents worked with offered upstairs laundry rooms located near the bedrooms, and buyers loved this convenience. He also gained other valuable input about some extras he was putting into his homes that he thought buyers would like. The agents told him that buyers in this market didn't care much about skylights or vaulted ceilings, due to energy costs.

Now Ken has a good handle on who and where his buyers are and their lifestyles and preferences. The only information he thinks he is lacking is how he should position himself against his competition and who that competition really is.

Know Your Competition

Who are your competitors? What products and services do they offer? How can you outsell them? Knowing what buyers want, although critical, is not enough. You also must figure out how to meet and exceed their expectations, and how to offer them what other builders are not.

When sizing up your competition, look at all builders, not just those of comparable size. Even if you are a custom builder, understand that even large national builders are now offering upscale options and customizable products. Consider all builders who might want a piece of your market share.

Consult Housing Market Analysts

A number of services in many metropolitan areas maintain comprehensive databases on new home communities in a given area. These services typically summarize trends, floorplans, price ranges, the builders who offer them, and other market information. These services are available through a subscription or a one-time fee. Consult your local home builders' association (HBA) to locate a reputable company that provides these services in your area.

Homes Valuation Websites

Several Internet sites compile data from a variety of sources that you can access for free. They are quick resources that provide recent home sale prices as well as the current price of homes still on the market. Some have mobile apps that allow you to quickly glance at a particular community or home to check what it sold for the last time it was on the market. These sites include:

- NewHomeSource.com
- Trulia.com
- Zillow.com
- Realtor.com

Tap into the Multiple Listing Service (MLS)

This wealth of information is available to members of your local Board of Realtors®. This information is similar to some of what you will find on the free home valuation sites above, but MLS is more accurate and up-to-the-minute current. As a builder you can join the local board as an associate member by paying an annual fee or you can rely on a Realtor® member for this information. These records show the properties that are currently listed for sale with real estate agents (both new and existing homes), their locations, prices, and square footages. In addition, the MLS shows homes recently sold within a specific time period and for what percentage of their listed selling prices.

Use Form 1.4 to record information obtained through the MLS or through a homes valuation website to get a summary of your competition, including resales. You can compile one list for properties currently on the market and another for those that have sold recently. Usually you will find a disparity between the listing and selling price of existing homes. That disparity will tell you a great deal about the current market demand for homes in your area.

Check Building Permits

Available from your municipality for free or a nominal fee, these public records will show all new home permits issued in a specific time period, the builder's name, and the location and size of the home being built. Use Form 1.5 to record your own findings.

Shop the Competition

Visit the other builders in your community to see what they offer. You may do this yourself, send a staff member to do it, or hire a mystery shopper. Drop by the sales offices, models, and inventory homes of all builders offering a product in a comparable price range. Gather brochures, floor plans, price lists, and a list of their standard features and upgrades. For each builder, make note of the following:

- Price range offered $____________ through $____________
- Detached or townhome
- Number of one-story plans offered
- Number of two-story plans offered
- Number of great room plans offered
- Number of dual master plans offered
- Bonus room offered?
- Guest house or carriage house offered?
- Square footage offered ____________ through ____________

Form 1.4 Competition Summary Report

Properties Currently For Sale on the Market

Address	Type of Dwelling (SF, MF, etc.)	Square Footage	Sales Price	Sales Price per Sq. Ft.

Properties Recently Sold

Address	Type of Dwelling (SF, MF, etc.)	Square Footage	Sales Price	Sales Price per Sq. Ft.

Form 1.5 Building Permit: Summary Report

Properties Currently For Sale

Address	Type of Dwelling (SF, Town-home, etc.)	Square Footage	Sales Price	Price per Sq. Ft.	Days on Market

Properties Recently Sold

Address	Type of Dwelling (SF, Town-home, etc.)	Square Footage	Sales Price	Price per Sq. Ft.	Days to Sell

- Detached or attached garage? Number of stalls? Tandem?
- Total number of floor plans ____

Summarize the information on each floor plan using the competitive analysis Form 1.6. Then you can easily compare what designs and square footages your competition is offering , which ones they are not, and how you measure up.

ANALYZING YOUR COMPETITORS' STANDARD AND OPTIONAL FEATURES

Keeping track of which features your competition offers as standard and which are priced as upgrades will help you stay competitive. If no one else in your market offers granite countertops as a standard feature, for example, here is where you can one-up your competitors. On the other hand, if everyone offers Energy Star appliances standard and you have been charging those as upgrades, maybe this is an area where you are losing to your competition. Use Form 1.7 to record your findings on what other area builders are offering. Note the additional cost.

Compile Your Results: The Competition

Once you have gained insight into what your competition is offering, the remaining challenge is using it to narrow your focus to your most likely competitors and identify strategies to outsell them.

Composite Builder: The Competition

In his competitive market analysis, Ken Smith, his wife, and his part-time office manager shopped the competition's sales offices, websites, models, and speculative homes. They obtained all the brochures, collateral materials, floor plans, and price lists they could from every builder offering homes between $125,000 and$500,000 in the southwest part of town.

They reviewed all the information they had compiled using Forms 1.6 and 1.7. Their analysis included a comparison of the features that other area builders were offering, and floor plan details such as the number of garage stalls, whether a bonus room was offered, and how much square footage was devoted to kitchens, family rooms, and laundry rooms. Ken discovered that even though the focus group agents told him buyers wanted computer niches, only a few area builders offered them. Only two builders offered oversized laundry rooms with sinks and folding counters. Three builders offered a third stall in the garage, but as an upgrade. One builder was offering a bonus room.

A review of the MLS listings showed that most of the new homes recently sold by real estate agents in the targeted price range were selling for a much lower price per square foot than he was offering. Building permits showed that homes in the southwest part of his city were being constructed at a crisp pace.

Armed with this information, Ken decided that he would eliminate vaulted ceilings and sky-lights to lower his cost per-square-foot and begin offering standard three-car garages and larger, more functional kitchens and laundry rooms. His next inventory home would show a computer niche adjacent to the children's rooms and he would vignette one of the bedrooms as a sewing and crafts room and tout it as a bonus room.

His competitive analysis led him to eliminate four of eight builders and focus on the remaining four as his likely competitors. Two offered products that were not as upscale as his, and two others targeted a retirement market with a patio home that had little or no yard.

Ken's next step, as well as yours, would be to allocate the funds for his marketing efforts.

Form 1.6 Competitive Analysis

Builder ______________________________

Community ______________________________

Model ______________________________

Price ______________________________

Size ______________________________

Price per Sq. Ft. $ ______________________________

Detached or townhome ______________________________

Stories ______________________________

Bedrms. ______________________________

Baths ______________________________

Size Master BR ______________________________

Dual Masters? ○ Yes ○ No

Size BR2 ______________________________

Size BR3 ______________________________

Size BR4 ______________________________

Study/Loft Size BR5 ______________________________

Size Library/Bonus Room/Media Room ______________________________

Size Game Room ______________________________

Size Closet Master BR ______________________________

Size Closet BR2 ______________________________

Size Closet BR3 ______________________________

Size Closet BR4 ______________________________

Size Closet BR5 ______________________________

Detached Guest house or carriage house? ______________________________

Size Linen Closet ______________________________

Size Coat Closet ______________________________

Size Great Room ______________________________

Size Family Room ______________________________

Size Living Room ______________________________

Size Dining Room ______________________________

Size Bkfst. Room ______________________________

Size Kitchen ______________________________

Detached or Attached Garage ______________________________

Size Garage ______________________________

# of Garage Stalls Tandem garage ○ Yes ○ No

Separate Laundry Room ○ Yes ○ No

Sink in Laundry ○ Yes ○ No

Basement ○ Yes ○ No

Finished ○ Yes ○ No

Size finished ______________________________

Size unfinished ______________________________

Vaulted ceilings ○ Yes ○ No

In what rooms and approximate heights:

______________________________ ft.

______________________________ ft.

______________________________ ft.

______________________________ ft.

Covered porch ○ Yes ○ No

Size ______________________________

Deck ○ Yes ○ No

Size ______________________________

Patio ○ Yes ○ No

Size ______________________________

Fireplace ○ Yes ○ No How Many? ______________________________

Kitchen Cabinets in linear ft. ______________________________

Upper ______________________________

Base ______________________________

Pantry ______________________________

Kitchen Island ○ Yes ○ No

Kitchen open to Fam. Rm. ○ Yes ○ No

Butler's pantry ______________________________

Master Suite Bath

Size ______________________________

Number of vanities ______________________________

Vanity space in linear ft. ______________________________

Tub ○ Yes ○ No

Jets ○ Yes ○ No

Sitting or exercise area ○ Yes ○ No

Private toilet room ○ Yes ○ No

Adapted courtesy of Quincy Johnson & Associates, Boca Raton, Florida

Form 1.7 Standard vs. Optional Features

Builder ______________________________

Use S for Standard, U for Upgrade, or N if not offered.

Item	Standard or Upgrade	Additonal Cost
Additional garage stall/tandem garage	______	______
Bonus room or game room	______	______
Computer niche	______	______
Deck/patio	______	______
In-ground swimming pool	______	______
Exterior hot tub	______	______
Fireplace(s)	______	______
Garage door opener/keyless entry	______	______
Kitchen island	______	______
Oversized showers in master bath	______	______
Patios, decks, and porches	______	______
Premium homesite	______	______
Recessed lighting or other lighting upgrade package	______	______
Recreational vehicle parking	______	______
Security system	______	______
Skylights	______	______
Vaulted ceiling in specified room(s)	______	______
Soaking tub in master bath	______	______
Stone or ceramic tile floors	______	______
Granite (or marble or other upgrade) countertop in kitchen	______	______
Stainless steel appliances in kitchen	______	______
Upgraded, energy-saving appliance package in kitchen	______	______
Granite (or marble or other upgrade) vanity in baths	______	______
Media centers, bookshelves, desks or other built-ins	______	______
Others that may be common in your marketplace:		
______	______	______
______	______	______
______	______	______
______	______	______

DETERMINE YOUR MARKETING BUDGET

2

After you have conducted your market research, know who your buyers are, understand what your competition is offering, have designed your product, and have determined its price range, you must establish a budget for the rest of your sales and marketing activities.

The Marketing Budget as a Percentage of Annual Sales

The most common method of allocating dollars to sales and marketing activities among small- to medium-volume builders is by percentage of projected gross annual sales (PGAS). Form 2.1 shows a simple formula for projecting annual sales. Your research should have given you an idea of the approximate number of homes being sold in the price range you want to target. Base your annual goal on how many of those sales you want to have the next year. Set it high enough to meet your profit goals but not higher than your company can reasonably achieve. Unrealistic goals are discouraging rather than motivating.

Form 2.1 Projected Gross Annual Sales

Annual Sales Goal (number of homes) $ ________

Average Price of Home You Will Offer $ ________

(multiply the two numbers)
Projected Gross Annual Sales (PGAS) $ ________

Sales and Marketing Budget Guidelines

Next, review the percentages in Table 2.1. These are typical ranges for builders. The actual percentage a builder spends depends on market conditions, geographic location, degree of competition, target buyers' shopping habits, and the builder's aggressiveness in the market.

Note that the budget guidelines are broken down into three categories:

- Essential to survive in business
- Necessary to maintain your market share
- Important to expand your reach

Table 2.1 Sales and Marketing Budget Guidelines (in Percentages of Gross Annual Sales)

Item	Percentage
To Survive	
Research	.25%
Sales	2.25% to 5.75%
To Maintain Your Image	
Online presence (includes website maintenance, social media efforts, online sales counselors)	.50% to 2.5%
Collateral material (stationery, brochures)	.25%
Model home and sales office (includes monthly payment, maintenance, phone, etc.).	.50% to 2.0%
Signs	.25% to .50%
Public relations	.25%
To Expand Your Reach	
Advertising: includes print, radio/TV, and online	1.5% to 2.5%
Special events (grand openings, promotions)	.25%
Total sales and marketing budget	.50% to 11.5%

Now apply the percentages to your projected annual sales figures, and you should have your sales and marketing budget breakdown.

Calculate Your Sales and Marketing Budget

Use Form 2.2 to record your own figures. The ranges you get after doing this exercise will provide some latitude when planning your marketing activities. The exact percentage you spend on advertising, on-site marketing, and merchandising activities will depend on local customs and conditions in your marketplace. In addition, you must continually monitor your sales and marketing efforts and measure the effectiveness of your overall marketing program to see where your budget may need to be periodically modified.

Form 2.2 Annual Sales & Marketing Budget Based on Projected Gross Annual Sales

Research:	PGAS of $________	× .25%	= ________		$ ________
Sales:	PGAS of $________	× .25% to 5.75%	= $ ________	to	$ ________
Online Presence:	PGAS of $________	×. 5% to 2.0%	= $ ________	to	$ ________
Collateral material:	PGAS of $ ________	× .25%	= ________		$ ________
Model Home, sales office:	PGAS of $________	× .25% to 2.0%	= $ ________	to	$ ________
Signs:	PGAS of $________	× .25% to .50%	= $ ________	to	$ ________
Public Relations:	PGAS of $________	× .25%	= ________		$ ________
Advertising:	PGAS of $ ________	× 1.5% to 2.5%	= $ ________	to	$ ________
Special events:	PGAS of $ ________	× .25%	= ________		$ ________
	Total Budgeted Sales and Marketing Expenditures				$ ________

Adapted with permission from McCurley and Sayre Residential, Inc. Pensacola, Florida.

Now you have an idea of how much of your total budget to spend on sales and marketing, and where to allocate those dollars, the next step is to develop a marketing strategy. This strategy will largely depend upon the market you are targeting. After your market research efforts, you should know whether your buyers are: first-time home buyers, move-up buyers, empty-nesters, or active adults and/or retirees. See the next chapter to get started on how to reach those specific groups and their submarkets.

DEVELOP YOUR MARKETING STRATEGY 3

Now you know who your prospects are: what they can afford, something about their lifestyles, characteristics, and preferences; and what your competition is offering. In addition, you have projected sales for the year and have established a sales and marketing budget. Now you must develop a marketing plan to spend those dollars most efficiently. To do that, you must

- narrow your focus;
- eliminate wasted efforts; and
- target your market.

Identify Your Target Market

This chapter discusses marketing strategies to employ to make sure your homes appeal to your targeted buyers. Not all of these will work for every location, but many will. In addition, once you train yourself to think in terms of targeted strategies, you may generate other ideas that will work in your specific market. These strategies are divided into three categories:

1. Products and designs to offer
2. *Merchandising* or *vignetting* your inventory homes or models
3. Advertising and promotion strategies

To merchandise is to fully furnish and decorate a builder's model or inventory home so it appeals to the intended target market. To vignette is to demonstrate the use of homes in inventory or model homes that are not fully furnished by using small decorator touches and accents (accessories) in such a way that the home will appeal to the intended target market.

From the information you gathered in your market research, you should be able to determine your target market. Buyers usually fit into one or more of four major buyer categories:

- First-time home buyers
- Move-up buyers
- Empty-nesters
- Active adults and/or retirees

First-Time Home Buyers

- Median age of 30[3]
- Satisfied with renting longer than in generations past
- Highly tech-savvy

- Environmentally conscious
- Driven by price and value
- Better educated and more sophisticated about the home buying process than first-timers in the past
- Want to live close to shopping, restaurants, entertainment centers, and, if they travel, a major airport
- May be planning to start or expand their families in the near future

First-Time Home Buyer Submarkets

The first-time market can include any of the following submarkets. Their distinct characteristics are discussed below.

MILLENNIALS

An emerging generation of home buyers of which home builders should be aware are the "Millennials," born after 1980 and growing up in the digital age. They are also called "Generation Y" and are the children of the baby boomers. This cohort will make up the future home buyer industry, so it will pay off to understand their needs and preferences, which are decidedly different from that of their parents. Below are some general characteristics to know about this vibrant group.

- More technologically proficient than any earlier generation
- Globally minded and comfortable with diverse populations
- Socially and environmentally aware; energy efficiency and recycled building materials will likely be important to them
- Likely to be patient with the rental market, waiting longer than their parents to buy their first home
- Fifty-nine percent said they prefer diversity in housing choices; 62% prefer developments offering a mix of shopping, dining and office space; and 76% place high value on walkability in communities[4]
- Prefer downtown locations near restaurants, entertainment, and public transportation
- Have a keen ability to see through hype and exaggeration and expect honesty and forthright communication.[5]

Millennial buyers may be singles, dual-income couples with no kids (DINKs), or young families. Below are some characteristics of each.

SINGLES

- Somewhat knowledgeable about the home-buying process
- Comfortable with technology
- Want to reap financial rewards with a home purchase, such as tax benefits and equity
- View their first home purchase as a stepping stone, rather than a long-term residence
- Concerned about security and privacy
- Want proximity to recreational activities, such as exercise facilities, pool, and jogging trails.

DUAL-INCOME NO KIDS (DINKS)

- Likely to have much more discretionary income than singles or young families the same age
- Are often the target of advertisers of luxury goods and travel opportunities
- Well-educated, professional, and knowledgeable about the home-buying process
- Comfortable with technology
- Usually entertain at home more than single professionals

YOUNG FAMILIES (SOMETIMES REFERRED TO AS DEWKS, DUAL EMPLOYMENT WITH KIDS)

- Seek a comfortable and attractive, but not ostentatious, home
- May be intimidated by too much glitz and glamour
- Express interest in current trends, but also tradition
- Occupied with leisure activities that revolve around their children and therefore see value in proximity to parks, good schools, and family entertainment.
- Concerned about child safety, so secure neighborhoods and home safety features are strong selling points
- When marketing to more experienced home buyers, consider not only the product offering but also how to merchandise and promote it.

Figure 3.1 Kids' Retreat

At McClelland's Creek, Ryland Homes of Denver transformed a loft into a kids' retreat, complete with built-in computer and homework space, cubbies and backpack storage. The design was developed with input from focus groups via social media sites such as Facebook, Twitter, Pinterest, and Houzz. Merchandising and social media design development by TRIO Environments. *Photo courtesy of Angela M. Harris, Trio Environments.*

Move-Up Buyers

- Mostly married couples with one to three children
- Looking for their second, third, or fourth home
- Need more room and want increased status
- Want a high degree of customization; even if they are purchasing a production home, they want ample opportunities to choose options and custom features
- May want to impress peers and colleagues
- Children may be older than those of first-time buyers so their space and room needs may change in the near future
- May be working at home
- Want flexible spaces that can meet changing needs, such as hobby rooms that can double as guest rooms
- Want space for entertaining since they are getting more free time as children grow up
- Want separate zones for different family members, such as a bonus/game room for the kids, a media room for Dad.
- Concerned with security issues

Checklist 3.1 Marketing Strategies for Move-Up Buyers

What to Offer Move-Up Buyers

___ Homes in gated communities and/or near walking paths

___ Security and climate systems that are smart phone controlled

___ Upscale finishes throughout, as well as granite (or similar) countertops, stainless steel appliances, hardwood floor throughout, high-end cabinets, millwork (such as fireplace mantels and coffered ceilings).

___ Upgraded lighting packages such as recessed lighting, can lights, and track lighting

___ Spacious master suites with large closets

___ Luxurious touches in the master bath, such as marble finishes, spacious walk-in closets, soaking tubs, separate vanities for his and hers

___ Impressive entries and foyers, such as curved staircases, full-volume ceilings, impressive chandeliers

___ Outdoor entertaining areas, such as outdoor kitchens, outdoor flatscreens, built-in BBQs

___ Built-in bookcases, media centers, art niches and work desks for artwork and collectibles

___ Spacious kitchens with islands and separate work zones

___ Kitchens open to family room

___ Pendant lighting over kitchen island

___ Warming drawers in kitchens

___ Wine chillers or built-in wine storage

___ Secondary kitchen in bonus room or on basement level (with small beverage refrigerator and microwave)

___ Home offices or spacious computer niches in the kitchen, loft, or master suite

___ Ample space for entertaining

___ High-technology systems such as smart systems, security systems, computer rooms

___ Media rooms with built-ins for flat-screen televisions, sound systems, built-in computer stations

___ Flexible rooms, for example, a hobby room or older child's room that can become a guest room as needed, or a bonus room that can be built as an option in the third stall of a three-car garage and used as a hobby or sewing room

___ Large walk-in closets in children's bedrooms

___ Ample opportunities to customize floor plans

___ If basements are customary in your market, offer guest suites or home offices with private, separate entries.

How to Merchandise or Vignette for the Move-Up Buyer

___ Use a touch of glitz and glamour in merchandising or vignetting.

___ Show entertaining in progress, such as wine bottle with wine glasses, a game room with a pool table, refreshments at a bar, cards or backgammon on a game table.

___ Show a coffee tray with a coffee press, two cups, and a newspaper in a master bedroom sitting area.

___ Show a bedroom as a guest room, home office, or exercise space.

___ Use dining rooms and media rooms to show entertainment possibilities.

___ Use warm, rich colors and finishes to fill up large spaces or rooms with high ceilings that may seem cold.

___ In furnishing large rooms, make them seem cozier and warmer by using large pieces such as a four-poster bed.

___ Show a bonus room with hobbies in progress, such as:

- a sewing room, complete with a sewing form, old-fashioned sewing machine, and shelves with ribbons, spools of thread, and fabric swatches.
- Create a pottery room with potter's wheel, colorful ceramics, and container of clay and glaze.
- For an artist room, set up an easel with tubes of paint. Paint the walls with splashes of color (over strippable wallpaper so you can easily remove the color later).

Advertising and Promotional Strategies for the Move-Up Buyer

___ Suggest a prestigious aura on your website and in ads. For example, show a luxury car parked in a circular drive.

___ Show family images on your website and in ads.

___ Emphasize proximity to good schools, parks, shopping, and recreational opportunities on your website and in ads.

___ Emphasize prestigious location. For example, show a country club lifestyle.

___ Offer a landscaping allowance. Give buyers the opportunity to work with your landscape architect to custom design their own yards and gardens.

___ Offer a year's membership at a local country club or recreation club as a special promotion.

Empty-Nesters

- Typically range in age from the mid-40s to 60s
- Want to reduce square footage and maintenance, but not status, price, or prestige
- Often have children who are away at college or are in various stages of moving in and out (a la boomerang children)
- Usually are established financially, may still be in the workforce, starting to plan for retirement
- Want to enjoy more leisure time and surround themselves with the finer amenities and possessions they could not easily afford during their child-rearing years
- May be caring for elderly parents now or in the near future
- Enjoy an active lifestyle, more than their parents did at the same age

Checklist 3.2 Marketing Strategies for Empty-Nesters

What to Offer Empty-Nesters

___ Single-story models

___ Outdoor entertaining areas, such as outdoor kitchens, built-in BBQs

___ Homes in gated communities

___ Spacious kitchens with islands and separate work zones

___ Kitchens open to family room

___ Pendant lighting over kitchen island

___ Warming drawers in kitchens

___ Wine chillers or built-in wine storage

___ Secondary kitchen in bonus room or on basement level (with small beverage refrigerator and microwave)

___ Some may desire patio or garden homes, with a minimum amount of grass to mow but still offering privacy and space to enjoy the outdoors

___ Formal dining rooms and open spaces for entertaining

___ Security and climate systems that are smart phone controlled

___ Two master suites to accommodate a live-in parent or returning child. A private entry is a plus via an above-garage apartment or quarters with a private entry on basement level

___ Separate guest house or carriage house

___ Three-car garages, parking pad for a recreational vehicle, or additional parking area

___ A first-floor master suite (if a multi-story product) with design options not feasible when they had children at home (for example, locating the master suite near the family room)

___ Built-in bookcases, media centers, art niches and work desks for artwork and collectibles

How to Merchandise or Vignette for Empty-Nesters

___ Show low-maintenance yards and lawns by demonstrating irrigation systems, rock gardens, smaller areas of grass (as appropriate for the climate).

___ Use garden walls and landscaping to make a strong first impression of a home that may be smaller than they are used to.

___ Demonstrate security systems.

___ Show secondary bedrooms as hobby or sewing rooms, offices, or dens.

___ Demonstrate a guest room or quarters for elderly parents.

___ Show entertaining in progress, such as wine bottle with wine glasses, a game room with a pool table, refreshments at a bar, cards or backgammon on a game table.

___ Show a coffee tray with a coffee press, two cups, and a newspaper in a master bedroom sitting area.

Advertising and Promotional Strategies for Empty-Nesters

___ Show an active lifestyle on website and in ads depicting recreational activities such as golf or tennis.

___ Show a strong sense of community on website and in ads, such as residents meeting in a gazebo or walking on a jogging trail in groups.

___ Emphasize low maintenance lifestyle in ads. For example, create an ad showing a resident mowing a large expanse of lawn with a "before" caption and a resident lying on a hammock near a rock garden with a waterfall with an "after" caption.

___ Offer a year's free service to a lawn and garden company as a special promotion.

Active Retirees

The "seniors" market is diverse, varied, and composed of several distinct submarkets. Age can range from 50+ and upwards. This diversity has been described as "from Eisenhower to Flower Power." In terms of housing, this group falls into one of these categories: active adult, age-restricted active adult, assisted living, memory care and long-term care. For the purposes of this book, we will address the active adult market, whether in an age-restricted or nonage-restricted community. Active retirees:

- May or may not have retired from the workforce. If so, they may have entered into a new, less-demanding career stage or semi-retirement.
- Are primarily finished with child-rearing, but an adult child may have moved back in; may entertain grandchildren frequently or occasionally
- Look for comfort and security
- Generally are affluent. Baby boomers (those born between 1946 and 1964) control 70% of disposable income in the U.S.[6]
- Perceive themselves as 10–15 years younger than they actually are (remember this when you are creating images for advertising)
- Are comfortable with new technology; boomers represent one-third of online and social media users.[7]
- Want to stay close to friends and family, especially grandchildren
- May be extremely financially cautious
- Usually have no urgency to move; they may have a strong emotional attachment to their present home. Need to be convinced that their lifestyle will improve substantially before they'll agree to move

Checklist 3.3 Marketing Strategies for Active Retirees

What to Offer Active Retirees

___ Homes in gated communities

___ Homes in communities that offer recreational opportunities such as walking trails, clubhouses, golf, swimming, tennis

___ Spacious kitchens with islands and separate work zones

___ Kitchens open to family room

___ Pendant lighting over kitchen island

___ Warming drawers in kitchens

___ Wine chillers or built-in wine storage

___ Secondary kitchen in bonus room or on basement level (with small beverage refrigerator and microwave)

___ Location near medical services

___ Low maintenance lawns, but do include spaces for outdoor entertaining, gardens and flower beds

___ Security system or package

___ Dramatic or bucolic views from bedroom, kitchen, and family rooms

___ Single-family, patio, or garden homes

___ Aging in place features, such as grab bars in tubs and showers (or the reinforcement in the wall so that they can be easily added). Consider scald-protection in kitchens and baths

___ One-story floor plans

___ Outdoor entertaining potential

How to Merchandise or Vignette for Active Retirees

___ Show a secondary bedroom as a visiting grandchild's room.

___ Avoid yellow tones in merchandising. As eyes age, they perceive colors with a more sallow cast.

___ Use bold, distinct colors in merchandising and decorating.

___ Use lighter colors, mirrors, and lighter tones of wood to maximize smaller floor plans.

___ Show wicker chairs, Adirondack chairs with wine and/or coffee serving on porches and decks.

___ Design all entries at grade (no walk-up entries).

___ Furniture in models should allow for ease in getting up and down.

___ Use flower beds and small gardens around the model, speculative home, or sales office.

Advertising and Promotional Strategies for Active Retirees

___ Feature people in ads, brochures, and graphics that are 10 to 15 years younger than the actual target market.

___ Emphasize an active, but leisurely, lifestyle in ads.

___ Emphasize security, peace of mind in ads.

___ In sales center or model displays, use large, easy-to-read copy on displays and collateral materials.

___ Furniture in the sales center should allow for ease in sitting down and getting up.

___ Walks and steps surrounding the model or speculative home, or for bigger communities, throughout the model center, should be obstacle free and nonslippery.

___ All areas used for sales purposes, including parking (if provided), the building entrance, and the internal area should be wheel-chair accessible. (This marketing strategy for retirees is also legally required for all sales centers by the American Disabilities Act.)

Other New Niche Markets

Builders should be aware of other emerging lifestyle trends that are creating new consumer needs. Listed below are some of the newest niche markets, their characteristics, and the strategies that might best appeal to each market.

Blended Families

In recent years, more and more couples have begun joining their lives and their families for the second—or more—time and creating new housing needs. Some characteristics of the increasing blended-family market follow:

- Family members are active and busy.
- The family includes a broad range of ages.
- Spouses have children from previous marriages. Some live in the home permanently and some may visit only on a regular basis.
- Both spouses are well established financially.
- Adult privacy is scarce and precious.

Checklist 3.4 Marketing Strategies for Blended Families

What to Offer Blended Families

___ Spacious kitchens with islands and separate work zones for multiple family members

___ Kitchens open to family room

___ Warming drawers in kitchens to accommodate different mealtime schedules.

___ Secondary kitchen in bonus room or on basement level (with small beverage refrigerator and microwave)

___ Give the parents private, romantic spaces that are separated from the kids' rooms.

___ Include computer niches outside of the children's rooms, or include them inside their rooms.

___ Give the children ample storage space, for instance, provide walk-in closets in the secondary bedrooms.

___ Large game rooms

How to Merchandise or Vignette for Blended Families

___ Merchandise a bedroom for a child who visits regularly. Accent the room with luggage or a backpack.

___ Target children's bedrooms toward a variety of life stages, for example, a nursery in one bedroom and a teenager's retreat in another.

___ Arrange game room for several activity zones, so occupants of multiple ages can engage in their own activities simultaneously. For example, place a Monopoly game in progress in one area and a video game on a flatscreen in another.

___ Put up a white board or use chalk board paint in the kitchen to create an activity control center with multiple messages and appointments scheduled. Hang up a soccer or swim-meet schedule on the refrigerator door.

The "Sandwich" Generation

Many people today find themselves in the "Sandwich Generation," a term coined by Carol Abaya in 2006. Between the ages of 40–65, they are named so because they are simultaneously caring for their children and their aging parents.[8] Some characteristics of these buyers are listed below:

- Generally, financially secure
- Retired or planning to retire in the near future
- Sophisticated and traditional tastes
- May prefer homes and yards that require little maintenance
- Active, busy lifestyle
- Generally reluctant to move, unless they can gain a great lifestyle benefit.

Checklist 3.5 Marketing Strategies for the Sandwich Generation

___ Merchandise one bedroom as an elder parent's room, or offer a separate apartment or guest house with a private courtyard or basement entry. (Be sure to check local zoning laws first.)

___ Merchandise a second bedroom as a teenage or college-age child's room; use pennants and memorabilia from a local community college or university.

___ Spacious kitchens with islands and separate work zones for multiple family members

___ Kitchens open to family room

___ Pendant lighting over kitchen island

___ Warming drawers in kitchens to accommodate different mealtime schedules

___ Secondary kitchen in bonus room or on basement level (with small beverage refrigerator and microwave)

___ Offer three-car garages and additional pads for parking.

___ Offer two master bedrooms.

___ Offer patio homes, small yards and gardens, irrigation systems, and/or low-maintenance landscaping.

___ Emphasize lifestyle benefits in your advertising.

Single Parents

Over 10.5 million Americans are rearing children on their own, according to the 2012 Statistical Abstract from the U.S. Census Bureau.[9] Understandably, most of them are seeking high-quality, affordable housing. Characteristics of this growing market follow:

- The home is empty during the day. The head of household works outside the home, and the child or children are at school and/or child care.
- The head of household, either male or female, is career-oriented.
- The single parent is time-conscious and has little free time for interior or exterior maintenance.
- The head of household is budget-conscious.

Checklist 3.6 Marketing Strategies for Single Parents

___ Provide microwaves at lower levels so that children can safely reach them.

___ Locate the kitchen next to the garage to make unloading groceries and other goods easier.

___ Locate the laundry room in close proximity to the bedrooms, upstairs if necessary. Install a pull-down ironing board.

___ Provide eat-in kitchens with bar seating at kitchen islands.

___ Merchandise the master suite in feminine colors and styles if you are targeting a single mom/with earth tones and masculine touches to target a single dad.

___ For a model targeted to a weekend dad, show a guest room ready and waiting for its occupant. Accent the room with luggage or a backpack.

___ Design open floor plans so the family room is visible from the kitchen so Mom or Dad can prepare dinner while watching the kids.

___ Use bright colors and contemporary styles and incorporate traditional touches.

___ Provide small, colorful gardens with a small amount of green space where the children can play, but not a lot to maintain.

___ For a feminine profile, show a relaxing bath in progress, complete with bath beads, candles, and a novel.

Boomerang Families

A residual effect of the Great Recession is the number of young adults living with their parents, as they complete school, pay off student loans, or find permanent employment. For example, young adults ages 25 to 29 were 17.9% more likely to live in their parents' households in 2010, compared to 2007.[10] The characteristics of this increasing market include:

- Multiple generations living under one roof.
- All family members are busy and on separate schedules.
- Privacy for all family members is at a premium.
- Storage is crucial.
- Space needs change frequently, flexibility spaces are important.

Checklist 3.7 Marketing Strategies for Boomerang Families

___ Design the master suite to be a romantic get-away: offer a luxurious bath, a fireplace, and a walk-out deck.

___ Provide all bedrooms with ample storage space, for instance, walk-in closets in the secondary bedrooms.

___ Offer a large game room merchandised to allow for several simultaneous activity zones. For example, exercise equipment in one corner, a computer niche in another, a flat-screen TV in another.

___ Merchandise a secondary bedroom for a young adult; decorate it with pennants from a local university or community college, stack some college textbooks on a built-in desk, and include a stair-climbing machine in the corner.

___ Put up a white board or use chalk board paint to create a memo board in the kitchen with multiple messages and appointments scheduled.

By now you should be getting a clear picture of how your research, your budget, and your target market will direct your sales and marketing compass. But before you embark any further on your journey you need to make some decisions about how your firm will operate. Chief among them is how you will staff your organization in terms of sales personnel.

STAFF FOR SALES SUCCESS 4

Now you are ready to make some important decisions about what will become one of your most important assets—the people who will have direct contact with your buyers. Whether these sales professionals will be real estate agents listing your homes or whether you hire one or more on-site sales representatives as employees, they will have the enormous responsibility of creating a positive first impression with your customers, communicating the benefits of your homes, and demonstrating how your homes are superior to other new or used homes. How your salespeople interact with your prospects can make or break your bottom line.

Determine Your Sales Staffing Requirements

You can structure the staffing of your sales personnel in three ways:

- Rely on the local real estate community.
- Employ on-site sales staff.
- Use a hybrid of outside brokers and on-site sales staff.

Each method has its benefits, and the best option for you will depend on the size of your company, your budget, your local market conditions, and your targeted buyers. The advantages and disadvantages of each are listed below to help you decide which structure is best for you.

Rely on the Local Real Estate Community

Advantages

- Lower sales office and sales management overhead
- Sales management and training handled by an outside source
- Little or no staff required
- Access to expertise a builder or builder's staff might not have; for example, financing, pricing, marketing, and familiarity with local media
- Good source of information about your competitors' products

Disadvantages

- Less control over sales management functions
- Sales staff may be inconsistent as agents from several firms may handle your listings and show your homes
- Less product knowledge of your homes
- Less loyalty to you (An agent may convert a prospective buyer to another builder or to a resale home if he or she is unsuccessful in the first sales attempt.)
- Sales staff may prefer to sell completed inventory homes rather than selling contracts to build because of the lag time in earning commissions

Employ On-site Sales Staff

Advantages

- Control over your sales training and management functions
- More consistency in staffing
- Better control over and more consistency in prospect follow-up
- Guaranteed exclusivity (The salesperson represents only your products.)
- Superior staff knowledge of your homes, benefits, and features, compared with an agent's

Disadvantages

- Added sales office and management overhead
- Limited customer base without access to a multiple listing service (MLS)
- Limited access to competitive information without MLS
- Lack of access to relocation networks for work-related transfers into your area
- Missed opportunities in the move-up buyer market. These buyers list their current homes with a broker while searching for a new home.
- Less exposure of your firm to the overall community
- Showing homes and managing sales staff is more difficult when you are offering homes in scattered sites.

Hybrid Approach of Outside Brokers and On-site Sales Staff

Advantages

- Reaps advantages of both methods
- Provides more control over training and management
- Access to a broader customer base, including clients of area brokers
- Divides the risks between builder and broker

Disadvantages

- May complicate tracking and payment of split commissions
- Increased commission payments in some cases (An on-site agent should be paid a consistent commission regardless of whether an outside broker is involved.)
- Requires a written policy to avoid disputes among the builder, on-site sales team, and outside brokers, that includes a commission agreement
- On-site staff must monitor and keep records of broker activity

Builder-Developer Sales Programs

Sometimes custom builders may purchase home sites in a community where the developer has its own sales and marketing staff, or they may contract with a new home sales and marketing brokerage company to handle on-site staffing and marketing for the entire community. Although this sales and marketing support drives traffic to the small-volume builder, the benefits may be costly. Each relationship between a builder and a developer is different, and customs and costs vary from region to region, so consider these questions before entering into an agreement with a developer.

Questions to Ask a Developer

1. What will the developer's responsibilities be in terms of marketing, advertising, and on-site sales?
2. What will my responsibilities be?
3. What costs will the developer incur for development fees, curbing, paving, and utilities?
4. What will my initial costs be? What will my costs be thereafter?
5. What kind of reputation does the developer have?
6. How fast have the homes in this community been selling? If it is just starting up, how well did homes sell in the developer's previous communities? How many sold in the first three months?
7. How much traffic does the community generate on a weekly basis, or if the community is brand-new, how much traffic did previous communities generate?
8. Is the community located near amenities that my target market wants, such as shopping, restaurants, good schools, recreation, and medical care? If so, which amenities?

Staffing with On-site Salespeople

If you decide to hire your own on-site sales staff, your staffing needs may fluctuate with the seasons or you may need temporary help during a special promotion that draws in extra traffic. If your traffic is or becomes higher than your team can manage, you will need to add new staff. Depending on the needs of your customers, not everyone on the sales team will need to be a sales representative.

Employ a Sales Assistant or Model Attendant

Some builders supplement their sales efforts by using a sales assistant in conjunction with their on-site sales representatives. A sales assistant does not require the same level of training as a full-time sales professional and is usually paid an hourly rate. (A good rule of thumb is to pay them $1 more per hour than a salesperson in a major retail store.) They can help out on the weekends when traffic is heaviest, during a special promotion that generates heavy traffic, and when sales associates are off for vacations or illnesses. The sales assistant can perform the following duties:

- Greet customers who arrive while the salesperson is busy with other guests.
- Register customers and engage them in conversation to learn more about their needs.
- Orient the customers to the company and community until the full-time salesperson is available.
- Assess the prospects' buying potential.

Choosing a Qualified Real Estate Agent

If you decide to list your inventory homes with a real estate agent, the questions in Form 4.1 will help you compare your choices and select the one most qualified—and the best choice for you.

Builder-Broker Cooperation

If you choose to employ your own on-site sales staff, you will still benefit from participating in the real estate community in your area. Full participation or cooperation means you welcome them to bring clients to you and you honor your commission agreements with those agents. Cooperating with the real estate community

Form 4.1 Choosing a Qualified Real Estate Agent

1. The listing agent a member in good standing of
   a. Local Board of Realtors® ○ Yes ○ No
   b. Multiple Listing Service (MLS) ○ Yes ○ No
   c. Relocation network with access to people whose companies are transferring them into the area ○ Yes ○ No
   (In areas that experience a great deal of in-migration, these cooperative networks should contribute at least 30 percent of your sales.)
2. Successful track record in selling new homes? ○ Yes ○ No
3. Willing to meet regularly with the builder to review progress and prospects, develop strategies, and solve any problems that arise? ○ Yes ○ No

   How often? ____________________
4. Marketing support the agent's firm offers

   ____ Online presence

   ____ Market research services

   ____ Advertising placement (print or online)

   Publication or website ____________ Frequency ____________

   Publication or website ____________ Frequency ____________

   Publication or website ____________ Frequency ____________

   ____ On-site signage, area directional signage

   ____ On-site collaterals

   ____ Open houses ____________ Frequency? ____________

   ____ Special events/real estate office tours ________ Type and frequency? ____________

   ____ Special education and certifications
5. Has the prospective listing agent earned either of the following:

   ____ Master in Residential Marketing (MIRM). This is the top designation from the NAHB Institute of Residential Marketing for specialized training in marketing new homes.

   ____ Certified New Home Sales Professional (CSP). Designees demonstrate a broad understanding of the home building business and advanced techniques for greeting, closing, and overcoming objections.
6. Will the listing agent be selling other builders' homes and/or resale homes as well? (A yes response should not necessarily eliminate a candidate, but you should know what to expect.)
   ○ Yes ○ No
7. Is the listing agent well-trained and knowledgeable about financing and contract issues?
   ○ Yes ○ No

provides many benefits, and if you properly execute and carry out your agreements, these advantages will outweigh any disadvantages. You want your customer base to be as wide as possible, and having a broker relations program is one way to widen it. If you properly court area real estate agents, they are likely to cheerfully bring clients to you. To cultivate real estate agents in your community, follow the guidelines in Checklist 4.1.

Checklist 4.1 Broker Cooperation

___ Develop a written broker cooperation policy that states the commissions you will pay on homes sold by a real estate agent who brings a client to you.

___ Register each real estate agent who visits your site. (Decide whether you will accept registrations by phone or if you will require each agent to make a personal visit with each buyer he or she wants to register.)

___ Adhere strictly to your broker cooperation policy so you establish a reputation in the community of keeping your commitments.

___ Join your area Board of Realtors® as an associate member.

___ Attend and sponsor Board of Realtors® meetings, luncheons, and other special events on a regular basis.

___ Visit area real estate offices regularly. Provide flyers about your homes or any promotions or incentives you may have planned. You don't need to make appointments to deliver them. All agents will have an incoming mailbox in which you can place your flyers, and you can talk with any agents who are present at the time of your visit.

___ Make appointments with the sales manager or principal broker of selected real estate firms to deliver presentations during their weekly sales meetings.

___ Develop a company newsletter to email to the real estate community. This newsletter can be as simple as a one-page flyer giving updates on your available homes, or a four-page newsletter created online to include third-party articles of interest. (Be sure to get permission from the publisher to reprint any material you use from another source.)

___ Invite area real estate agents to your site by hosting a grand opening, luncheon, or special preview of your model or speculative home exclusively for agents.

___ Offer incentives such as contests or a drawing. For example, offer a free lunch, weekend getaway, or other giveaway to a real estate agent who sells one of your homes during a certain promotional period. (Check local and state gaming laws before offering any drawing or promotion.)

___ When a real estate agent sells one of your homes, hand-deliver a small gift or gift certificate to the agent with a thank-you card attached.

___ Offer a high-quality product at a reasonable price and provide outstanding customer service and warranty follow-up. Real estate agents will continue to refer clients to builders they know have a strong reputation.

Finding and Hiring On-site Sales Staff

Where do you find potential new talent? You should be continually searching for candidates for new home salespeople. Even when you are fully staffed, you may have a sudden opening that calls for quick action. Keep a file on hand of resumes or business cards of people you regard as good candidates. You can also look in the places referenced in Checklist 4.2 Finding and Hiring Sales Staff.

Checklist 4.2 Locating Candidates for Your Sales Staff

___ Local department stores with a high level of customer service, such as Nordstrom, where you are impressed with a salesperson's talents

___ Car dealerships or another business where you are impressed with the salesperson

___ Your competitors' sales offices

___ Area real estate agents

___ Professional recruitment agencies, particularly those who specialize in staffing builders' sales offices

___ Craigslist.com

___ Your local home builders' association

___ Your local Sales and Marketing Council

___ LinkedIn

Skills and Qualities of a Successful Salesperson

You can train just about anybody on the correct methods of selling and teach them the benefits and features of your new homes, but to find a true salesperson superstar, those traits must be inherent in the person you are hiring. To find a superstar salesperson, look for the qualities in Checklist 4.3.

Checklist 4.3 Superstar Salesperson Qualities

___ Demonstrated interpersonal skills

___ Excellent verbal and written communication skills

___ Proficient with latest technological tools

___ Excellent time management skills

___ Strong teamwork skills

___ Inquisitive, interested in learning

___ Excellent phone skills

___ Neat, professional appearance

___ High energy level

___ Enthusiastic

___ Positive attitude

Screen and Interview Candidates

Salespeople must first be able to sell themselves before they can effectively sell your homes.

Use Form 4.2, Sample Interview Questions and Assessment Criteria for guidance in determining who is best qualified to be on your sales staff.

New Salesperson Orientation

Immediately after hiring a new salesperson, provide a complete orientation to your company, including the information in Checklist 4.4.

Tip from a Pro

Melinda Brody, MIRM, recommends the following to ensure that your new-hires are as proficient with today's latest technology as your consumers are:

1. During the interview, ask the candidate to go out on site, take three photos of available homes, and text them back to you.
2. Ask if they can conduct an interview via Skype.
3. Find out if they are active users of social media.

Checklist 4.4 New Salesperson Orientation

Give your new-hires a thorough introduction of the following:

___ Company mission statement

___ Company history

___ Organization

___ Current projects

___ Company policies and procedures

___ Your company's website

___ The technological tools available to them

___ The community or communities' benefits and features

___ Your homes' benefits and features

___ Surrounding areas, including location of shopping, schools, parks, churches, and average commute times to job centers

___ Any social media sites that your company uses

Form 4.2 Sample Interview Questions and Assessment Criteria

Applicant's Name ____________________ Phone Number ____________________

Email address ____________________

1. Review resume, including past 5 years' history, responsibilities, and accomplishments. Write notes on resume.
2. What did you like best about your last (or present) company? ____________________
3. What did you like least about your last (or present) company? ____________________
4. Why did you leave your last company (or why are you looking elsewhere if you are presently employed)? ____________________
5. What has your gross income been for each of the past three years? $ ______ $ ______ $ ______
6. How much did you earn your best year ever? $ ____________ Year ______
7. Describe the ideal position for you. ____________________
8. What are your strongest competencies? ____________________
9. What are your greatest weaknesses? ____________________
10. What do you do better than other salespeople you know? ____________________
11. What personal improvement goals have you set for this year? ____________________
12. Where do you see yourself professionally in 3 years? ____________________
13. How do you feel about working evenings and weekends? ____________________
14. Tell me what you know about our company. ____________________
15. How can you contribute to our company's success? ____________________

Form 4.2 Sample Interview Questions and Assessment Criteria (*continued*)

Interviewer, please rate candidate on each of the following attributes using a scale of 1 to 10:

___ Presented professional, complete resume

___ Provided at least three business references that checked out favorably

___ Left professional voice mail message

___ Promptly returned phone calls or emails

___ Asked pertinent questions

___ Gave spontaneous responses

___ Has professional phone greeting

___ Was friendly

___ Stayed focused

___ Attempted to set follow-up appointment

Interviewer's Comments and Recommendations: ______________________________

____

____

____

____

____

____

____

____

____

____

Adapted with permission from Roger Fiehn & Associates, Houston, TX.

The Sales Manager's Role

If your company is large enough to have a sales manager to oversee your sales staff, he or she should bear the responsibilities found in Checklist 4.5. In smaller companies, the builder may have to be responsible for them.

Checklist 4.5 Sales Manager's Responsibilities

___ Develop or update a policy and procedures manual for sales functions.

___ Recruit and hire new sales representatives.

___ Conduct new salesperson orientations.

___ Establish personal sales goals for each salesperson.

___ Spend time with each salesperson on-site, in the sales office, and in the field.

___ Train and motivate the sales staff.

___ Monitor and evaluate each salesperson's performance.

___ Provide feedback to the salesperson regarding his or her performance.

___ Conduct weekly sales meetings to review goals and progress.

___ Provide ongoing training sessions.

Using Outside Consultants for Training

For small-volume builders who have only one or two salespeople, as well as large firms that simply want to provide a fresh perspective, outside sales training consultants offer services ranging from half-day motivational seminars to intensive training programs. To find a consultant, use the following sources:

- NAHB's Directory of Professionals with Home Building Designations found at http://tinyurl.com/knb9edq.
- Your local home builders' association or local Sales and Marketing Council (SMC). Check with your association to see if it has such a council. If you are not affiliated with a local association, you can find the nearest one to you by visiting NAHB website (nahb.org) and clicking on the "Find: Local Association" link in the upper right corner.

Retaining Successful Salespeople

Once you have hired and trained your sales staff, keep your best salespeople from being wooed away by your competition. The practices in Checklist 4.6 will help motivate your staff to remain productive and lucrative.

Checklist 4.6 Salesperson Retention

___ Provide ongoing training opportunities

___ Offer performance incentives, such as bonuses, free travel, or other rewards

___ Stage in-house sales contests periodically

___ Give positive reinforcement (from high-level company officials, such as the sales manager, builder and, in large firms, the marketing director and president) through handwritten notes or phone calls praising a staff member for a job well done

___ Recognition through NSMC's Million Dollar Circle Award (see nahb.org)

___ Regular recognition through a Salesperson of the Month or other award

___ Public recognition through the media (by press release, for example) for new hires, anniversaries, million dollar sales marks, winners of in-house contests, etc.

___ Regular and frequent reviews with the sales manager or builder to praise good performance, improve weak skills, and set new goals

Compensating Your On-site Sales Staff

You have several options in structuring your on-site sales staff's compensation package. The one that works best for your company depends on what is customary in your market, your annual sales volume, and the number of people you employ. Remember that your compensation program must be lucrative enough to motivate your sales team and compete with what other builders in your area offer. Following are three common approaches to compensation, followed by explanations of their advantages and disadvantages.

Straight Commission or Draw against Commission

Straight commission pays only when a home is sold, typically 2% of sale price. No health insurance, pension plan or other benefits. The salesperson is often designated as an independent contractor for IRS tax purposes. Often, in the beginning of employment, a salesperson is given a monthly advance against which his commissions are debited (called a draw) as they are earned.

Advantages

- Commission or draw is easy to calculate, track, and administer.
- Builder has the option of using independent contractors as opposed to hiring employees.
- Compensation is directly linked to performance.

Disadvantages

- On straight commission, salespeople experience a lag before payments begin.
- On a draw, salespeople may accumulate a deficit which may discourage motivation.

Commission Plus Benefits

Commission is calculated on a per-house-sold basis, typically 2%, but builder also offers health insurance and possibly 401(k) plan.

Advantages

- May attract more experienced and more skilled sales professionals.
- Provides plenty of incentive for the salesperson and gives him or her the security that comes with benefits, such as health insurance.

Disadvantages

- Benefits and expenses may be harder to administer.
- Builder must stay current on state and federal regulations governing employee benefits.
- May be too costly and time-consuming for a small-volume builder.

Straight Salary, Salary Plus Bonus, or Salary Plus Commission

Builders could choose to pay their salespeople a straight salary; a salary with incentives, such as bonuses; or a salary plus a commission per house sold.

Advantages

- Provides a secure base of income while still motivating the employee.
- May be good for a slow market or distressed property.
- May work well for new-hires; they can switch to straight commission as they become more experienced.

Disadvantage

- More expensive for the builder when sales are slower.
- May not provide as strong an incentive for the salesperson as a plan based strictly on performance.

Evaluating Your On-site Sales Staff

Two ways to evaluate your staff are by using a mystery shopper and conducting surveys to get feedback from customers and prospects.

Mystery Shopping

This method of evaluating an on-site salesperson's performance entails hiring a professional evaluator (the mystery shopper) to pose as a customer and visit your sales office in order to evaluate his or her performance. It is usually audio or videotaped with a hidden camera. Have your salespeople sign an agreement upon being hired that allows the builder to film and evaluate them in this way, so that they have full disclosure that this may happen at any time. This will also provide incentive to perform at their best during every customer contact.

If you use a mystery shopper, keep the following guidelines in mind:

- The shopping process is for training, learning, and improving your salespeople's on-site selling skills. It is not to catch someone performing badly.
- "Shop" each salesperson approximately every three months to measure progress.

- Make certain that your shoppers closely match the targeted profile of your prospects or they will not be credible.
- Reinforce your shopping program by offering rewards for positive shopping reports, for instance, a cash bonus or gift certificate.
- Protect yourself legally by obtaining the permission form described above.

For professional *mystery shopping* services, contact the NAHB Directory of Professionals with Home Building Designations at http://tinyurl.com/kn9edq. These firms will provide the shoppers, send you a detailed report and usually provide some training services as well. See Form 4.3 for a sample evaluation form for a mystery shopper to use.

Post Move-In Customer Surveys

Another way to evaluate your salespeople's performance is by asking your customers their opinions in a survey immediately after move-in. A sample questionnaire for such a survey appears in Chapter 13, Form 13.2, Sample Customer Service Survey.

Develop Strong Selling Skills

Today's successful on-site salespeople must hone the following skills and techniques to build rapport and gain a commitment from buyers.

Critical Path of Sales

Learn to follow the critical path of sales and the techniques that are associated with each step. Each of those steps is listed below with its applicable techniques.

GREET PROSPECTS

- Extend a proper greeting to all prospects: stand, smile, make eye contact.
- Obtain their names, and remember to use them often. (But not so often that you irritate the customer.)

QUALIFY PROSPECTS

- Enter every guest into your Customer Relationship Management program. Ask qualifying questions in a conversational manner as you enter the information.
- Establish a rapport, demonstrate empathy for prospects, and show strong listening skills.
- Pay attention to each customer's needs and take notes.
- Meet regularly with a mortgage representative to keep abreast of the most current financing tools and products so you can have an initial discussion with your prospects about financing.

DEMONSTRATE PRODUCT

- Demonstrate inventory home or model. If you are selling prior to construction, show floor plans, elevations and portfolio. An iPad or other tablet is a great way to do this. Go out into the community and visit the available homesites.
- Point out specific benefits and how this builder's homes will satisfy their needs.

Form 4.3 Sample Salesperson Evaluation

Sales Representative ______________________ Date ______________________

Community ______________________ Day of Week/Time ______________________

Please respond Yes, Somewhat, No, or Not Applicable:	Y	S	N	NA

Approach/Introduction

A. Did the sales representative:

1. Greet you immediately as you entered?	Y	S	N	NA
2. Introduce him/herself?	Y	S	N	NA
3. Ask your name?	Y	S	N	NA
4. Welcome you warmly and with enthusiasm?	Y	S	N	NA
5. Dress in a professional manner?	Y	S	N	NA
6. Register you at any time during your visit?	Y	S	N	NA

Comments ______________________

Qualifying/Needs Analysis

B. At any time during your visit did the sales representative:

1. Attempt to build rapport with you? In what way? ______________________	Y	S	N	NA
2. Determine your moving time frame?	Y	S	N	NA
3. Ask about your preferences in a home?	Y	S	N	NA
4. Ask what you have seen so far?	Y	S	N	NA
a. Ask about your likes and dislikes about what you've seen?	Y	S	N	NA
b. Ask how long you've been looking?	Y	S	N	NA
5. Ask why you are considering a move?	Y	S	N	NA
6. Ask how you heard about the community?	Y	S	N	NA
7. Build trust by listening to you?	Y	S	N	NA
8. Ask about the size of your household?	Y	S	N	NA
9. Discuss price range of homes?	Y	S	N	NA
10. Determine your knowledge of the builder?	Y	S	N	NA
11. Determine your knowledge of the area?	Y	S	N	NA
12. Ask if you currently own a home or rent?	Y	S	N	NA
13. Ask about your employment status/occupation?	Y	S	N	NA

Comments ______________________

Form 4.3 Sample Salesperson Evaluation (*continued*)

Presentation

C. Did the sales representative:

1. Offer to show you the models?	Y	S	N	NA
2. Ask you to tour models alone?	Y	S	N	NA
3. Personalize the demonstration to your specific needs?	Y	S	N	NA
In what way? ________________				
4. Talk in terms of benefits as well as features?	Y	S	N	NA
5. Discuss quality of construction?	Y	S	N	NA
6. Build perceived value?	Y	S	N	NA
In what way? ________________				
7. Sell you on the builder?	Y	S	N	NA
a. Sell you on the builder's unique advantages over the competition?	Y	S	N	NA
b. Offer information on the warranty program?	Y	S	N	NA
8. Sell you on the community?	Y	S	N	NA
9. Seem knowledgeable about the homes' construction?	Y	S	N	NA
10. Narrow you down to a specific model?	Y	S	N	NA
11. Use plot map to show available home sites?	Y	S	N	NA
12. Narrow your interest to one home site?	Y	S	N	NA
13. Tour home sites with you?	Y	S	N	NA
14. Ask you to go alone to available sites?	Y	S	N	NA
15. Sell you on the community's amenities?	Y	S	N	NA
How? ________________				
16. Volunteer information on any of the following?	Y	S	N	NA
__ Nearby shopping				
__ Schools				
__ Libraries				
__ Quality of Residents				
__ Recreation				
__ Churches/Synagogues				
__ Other				
17. Ask trial closing question during demonstration?	Y	S	N	NA
18. Seem knowledgeable about options?	Y	S	N	NA
19. Handle objections effectively?	Y	S	N	NA

a. Objection 1 ________________

Salesperson's Response ________________

b. Objection 2 ________________

Salesperson's Response ________________

Form 4.3 Sample Salesperson Evaluation (*continued*)

20. Maintain control of sales process during model tour?	Y	S	N	NA
21. Have overall product knowledge?	Y	S	N	NA

Comments ______________________________

Financing

D. Did the sales representative:

1. Determine your ability to purchase?	Y	S	N	NA
2. Initiate financing discussion?	Y	S	N	NA
3. Seem well informed about financing?	Y	S	N	NA
4. Offer to do a financial work-up?	Y	S	N	NA
5. Discuss initial investment?	Y	S	N	NA
6. Discuss monthly investment amount you felt comfortable with?	Y	S	N	NA
7. Discuss closing costs?	Y	S	N	NA
8. Discuss financing plans available?	Y	S	N	NA
9. Write down information for you?	Y	S	N	NA
10. Give you a work-up to take with you?	Y	S	N	NA
11. Sell financing as well as the home?	Y	S	N	NA
12. Ask you to contact mortgage company and/or bank directly for financial discussions?	Y	S	N	NA

Comments ______________________________

Close/Commitment

E. Did the sales representative:

1. Create a sense of urgency?	Y	S	N	NA
2. Fully explain purchase procedure?	Y	S	N	NA
3. Show you a purchase agreement?	Y	S	N	NA
4. Ask directly for a deposit?	Y	S	N	NA
5. Ask for a lot reservation?	Y	S	N	NA
6. Ask directly for a firm, future appointment?	Y	S	N	NA
7. Say he or she would follow up with a phone call?	Y	S	N	NA

Comments ______________________________

Form 4.3 Sample Salesperson Evaluation (*continued*)

Sales Representative

F. Was the sales representative

1. Energetic?	Y	S	N	NA
2. Sincere?	Y	S	N	NA
3. Well organized during the presentation?	Y	S	N	NA
4. Seemingly in control of the process?	Y	S	N	NA
5. A good listener?	Y	S	N	NA
6. Able to handle traffic well?	Y	S	N	NA
7. Informative?	Y	S	N	NA
8. Friendly?	Y	S	N	NA
9. Patient?	Y	S	N	NA
10. Genuinely interested in you?	Y	S	N	NA
11. Someone you would buy a home from?	Y	S	N	NA

If no, why not? ____________________

Comments ____________________

Signage/Community

G. Were directional signs noticeable as you approached?	Y	S	N	NA
H. Were signs well placed and readable?	Y	S	N	NA
I. Did the community give you a good first impression?	Y	S	N	NA
J. Were the models well kept, lighted, and climate controlled?	Y	S	N	NA
K. Were the models attractive and appealing?	Y	S	N	NA

Comments ____________________

Overall Comments ____________________

____

Positives ____________________

____

Items That Need Improvement ____________________

____

- Recognize and adjust your presentation to various personality types.
- Involve the potential customer's children (if any) in the presentation, and include them in the conversation.

OVERCOME OBJECTIONS

- Anticipate the most commonly presented objections, such as "ABC Builders is cheaper," and have a prepared answer, such as "Our homes give customers the best value, as they all include upgraded name-band appliances and windows as a standard feature." Note: Never disparage the competition.

CLOSE THE SALE

- Effectively handle any objections that arise.
- Create urgency or help to shorten the decision-making process.
- Ask for the sale.

FOLLOW-UP

- After a sale, reduce or eliminate buyers' remorse by contacting the customers at regular intervals during the building process, keeping them apprised of construction phases, financing approval, and completion date. Always reinforce how excited you are for them at each stage.
- Attend the mortgage application meeting with the buyer, and act as the liaison between the loan officer and buyer. This will help you to maintain control over the process.
- Enter customers and prospects into a Customer Relationship Management (CRM) program that will remind you when and how often to follow-up.
- Contact prospects and customers at regular intervals.

Network for Referrals

Salespeople should continually network for referrals from the following sources:

- Previous customers
- Prospects
- Local merchants
- Real estate agents
- Professional organizations
- Major employers

Establish a Prospect Management and Follow-Up System

The most successful salespeople are those who maintain a regular follow-up program for prospects. Listed below are the tools you need to develop and maintain such a program.

Customer Relationship Management System

Register every visitor to your corporate office, sales office, model or inventory home by entering them in a Customer Relationship Management (CRM) program. This database will form the basis of your prospect management program and will greatly assist in market research.

A number of cloud-based or automated follow-up systems can maintain a database of prospects, create a schedule for follow-up, and generate correspondence (already written for you) on a daily, weekly, and monthly basis. You can adapt these systems for your particular firm and prospects. To find a CRM program right for you, research the following resources:

- NAHB's Business Management & Information Technology Committee. This committee offers an online resource called BizTools.
- NAHB's Directory of Professionals with Home Building Designations at http://tinyurl.com/knb9edq.
- Ebuild.com. An online catalog database maintained by publisher Hanley-Wood. Search for Sales and Marketing.

You can also use a CRM program to conduct customer and prospect surveys. See Chapter 13, Form 13.2, Sample Customer Service Survey, for questions to include.

Prioritize Your Prospects

This simple strategy will help your salespeople prioritize their time and concentrate on those prospects most likely to buy. Classify each customer as an A, B, C, or D contact.

- "A" contacts are ready, willing, and able to buy. Concentrate particularly on these prospects, and follow up with them most frequently.
- "B" contacts are ready and able to buy, but they still need encouragement before they will become willing to buy.
- "C" contacts are ready and willing to buy, but they may not qualify financially. You still need to keep in touch with them regularly because their financial situations may change.
- "D" contacts are not likely to buy within the next year and are not qualified, but you still want to send one to three letters to them throughout the year. They may refer friends or colleagues your way and may eventually become buyers.

Prospect Follow-Up

As you continue to build your database of prospects, complete the steps in Checklist 4.7 for an effective follow-up strategy.

Checklist 4.7 Prospect Follow-Up

___ Call or email all visitors to your site immediately after their visits and provide answers to any questions they may have.

___ Send a handwritten thank-you card or postcard to prospects 24 to 48 hours after their initial visit. Thank them for their time and, again, invite them to call you with any further questions.

___ Guided by your CRM program, set up a "drip campaign," whereby periodic emails or letters go to each prospect, the frequency depending on the prospects' classification as A, B, C or D contacts. Include in the messages the following newsworthy items.

- Updates on phase openings
- Special on-site events

- Special pricing, promotions, or buyer incentives
- Company newsletter
- Copies of related newspaper clippings or press releases
- Any other information that may be of interest

___ Continue to follow up with every contact until he or she buys a home from you, or you are certain he or she is not going to.

___ After a prospect becomes a customer, continue to follow up at these stages:

- Financing approval
- Color and/or decorator selections
- Production schedule updates
- Change order status
- Prior to settlement
- Prior to customer orientation
- Immediately after move-in
- 30 days after move-in
- 90 days after move-in
- Just before warranty expiration date

Education and Certification for On-site Salespeople

The following educational programs are available from the NAHB and upon completion, qualify the participant for a certification/designation that will set them apart from their counterparts. For information on these programs, go to nahb.org, go to the Education & Events tab and click on "Designation Overviews and Resources."

- Certified New Home Sales Professional (CSP)
- Master Certified New Home Sales Professional (MCSP)
- Certified New Home Marketing Professional (CMP)
- Master in Residential Marketing (MIRM)

Now you should be able to make some well-informed decisions about your sales staffing. The next chapter looks at some important considerations about your silent salespeople—your logo, signs, graphics, and ads. All these elements—and more—will shape your company image in the public's eye. Turn to Chapter 5 to check your image.

CREATE YOUR COMPANY IMAGE 5

When prospects visit your website or drive to your sales office, inventory home, model home, or new home community, what impression do they form? How recently have you checked that impression? What kind of image do you portray to the community at large through your advertising and promotional activities? Regardless of your annual volume or price range, you can take simple steps to ensure that all elements of your sales and marketing program consistently project a first-class message of quality. Some critical considerations follow.

Lead with a Logo

The first step in building an image for your homebuilding company is to create a logo (or redesign your current logo) to express your quality and style. Use a professional graphic designer. You can find one through an advertising or public relations agency or an online service such as 99designs.com. The more information you provide about your company mission, your products and services, the more likely you are to get the logo you envision. Checklist 5.1 contains guidelines for designing an effective logo.

Budget-Saving Tip

To get a creative professional look for your new logo inexpensively, contact the art department of a local college or university. A class, a professor, or an advanced student may be willing to take on the design of your logo as a special project.

Checklist 5.1 Logo Design

___ Choose a color that will convey a strong image and be visible on signs and graphics. Ideally, these same colors will carry through in other branding efforts.

___ Design a logo that will reproduce equally well in color and in black and white.

___ Select a design and typeface consistent with your architectural style and target market. For example, a fun logo could appeal to a young family whereas an elegant, classic symbol would be appropriate for a high-end, move-up customer.

___ Keep the typeface simple enough that your name is readable. Avoid overly flowery lettering or elaborate type.

___ Include your logo on your website, social media, signs, ads, and banners.

___ Favor horizontal designs over stacked logos; they are easier to read.

Figure 5.1 Strong Builder Logo

Pictured is the logo for Eagle Construction, the winner of the Best Print Campaign category in the 2013 Nationals awards, the annual sales and marketing competition for home builders sponsored by NAHB's National Sales and Marketing Council. Copy is easily readable, there is adequate contrast between letters and background, and the logo suggests an image of strength and durability. The logo reproduces well on all of the builder's advertising and collaterals. To browse other winning designs from NAHB's Nationals Awards, see thenationals.com.

Develop an Effective Website

Research shows that the vast majority of buyers are now turning to the Internet in order to gather information and narrow their choices prior to visiting builders' model homes. As mentioned earlier, a 2001 study showed that 42% of home buyers used the Internet as an information source, but by 2013, that number had increased to 90%.[11]

Although some software companies offer programs that allow a layperson to design his or her own website, this strategy is not recommended for today's competitive builder. With technology becoming increasingly sophisticated and builder web design growing ever more competitive, you need to hire a web development company. Your advertising agency can be a good resource, especially if they specialize in real estate. Many agencies have added web design to their menu of expertise and services. Checklist 5.2 lists what elements an effective website should include.

Checklist 5.2 Effective Website Features

___ Include icons of all social media you use, and links to them ___

___ Utilize responsive web design that will adapt to different screen sizes of users

___ Employ *search engine optimization* (SEO) that helps search engines like Google find your website when users search on certain key words or phrases, such as "Dallas new homes"

___ Photographs of the homes, elevations, and interiors

___ The ability to sign up for your company or community newsletter, if you have one

___ Community information

___ Floor plans

___ Site maps that indicate available homes

___ A contact link to use for sending inquiries or requests for more information

___ Options and upgrades offered

___ Driving directions to your communities and homes

___ Tracking software

To properly market your website and generate traffic to the site, consider the following:

___ Include your website address is on all collateral materials and ads.

___ Fill it with interesting and colorful content.

___ Maintain a *blog* and make entries at least weekly.

___ Participate in social media, such as Facebook and Twitter.

___ Conduct email marketing campaigns.

___ Comment on other industry blogs.

___ Post high-quality videos on YouTube of customer testimonials, virtual home tours, showcases of neighborhoods, and special event coverage.

Successful Signage

Signs on and off your site will do a great deal to direct prospects to your site and project your image. You can take the steps in Checklist 5.3 to ensure that all the signs in your signage package make a strong, positive impact.

Checklist 5.3 Signage Package

___ Prominently display the builder's name and logo.

___ Include the builder's or sales office phone number and/or cell number.

___ Include builder website address.

___ Keep color, style, and typeface consistent on all signs and graphics.

___ Choose black or another bold text color that will contrast well with background color.

___ Periodically check all signs to see that they are well-maintained with no missing letters or chipped paint.

___ Replace signs when worn or faded.

___ Make signs large enough and the text bold enough to be seen from a passing car.

Listed below are the types of special signs you may need along with their characteristics.

Directional Signage

- Position these signs near major neighboring intersections in a visible location. Include an arrow directing the way to your home(s).
- Check on local ordinances for placement restrictions. Some require directional signage to be placed only on weekends or during certain hours.

Entry Signage

- Complement your entry sign or wall with well-maintained landscaping.
- A water feature is an attention-getting element in an upscale community.
- Use materials that are commensurate with the price range of your homes; for example, stone or brick in an upscale community.

Billboards, Bench Boards, Bus Boards, Mass transit

- Make graphics and headlines immediately attention-getting.
- Convey a customer benefit of your homes.
- Include phone number and website address.
- Headline in ad should be short and concise, but deliver a strong message.
- Include the location of your homes, sales office, or site.
- Include the price range of your homes.

Model Home Signs

- Identify each model with a distinguishing name; for example, The Wisteria, not Plan A.
- Place these signs where visitors can plainly see them, either on a model's front door or near the entrance to it.

Other On-site Signs

- Prominently identify homes or sites that are sold with a "Sold" sign. Personalize the experience by adding: "Future Home of Mr. and Mrs. Home Buyer."
- Prominently identify available inventory homes or home sites with a "For Sale" or "Available" sign that includes phone numbers and web address.
- For large communities, include directional signs to the clubhouse, community pool, tot lot, fitness center, golf course, jogging trail, bus or transit stop, equestrian center, etc.
- In a master-planned community, have street signs and traffic signs manufactured with material, color, and design consistent with other on-site signage (as applicable regulations allow).

Perfect Parking Lots

Give your prospects a smooth and flawless arrival at your site. Refer to Checklist 5.4 to make sure your parking lots welcome your guests appropriately.

Checklist 5.4 Parking Lots

___ Provide handicapped-accessible parking with obstacle-free ramps all the way to the speculative home, model home, or sales office. This accessibility is required by law.

___ Provide ample off-street parking.

___ Make sure the area is paved or graveled and free of mud, snow, or excessive dust.

___ If you are conducting a grand opening, a luncheon for real estate agents, or another special event, arrange for a parking attendant or valet parking.

___ Locate the parking area close to your inventory homes, model homes, or sales office.

___ Provide paved walks to and from sales office and to models.

Excellent Exteriors

Whether you have a sales office and/or model home complex or whether one inventory home does both jobs, the exterior should cast a strong first impression on your prospects. Your prospects should feel at home even before they enter the front door. Listed in Checklist 5.5 below are some tips for presenting your inventory home, model home, or sales office in its best light. (For more information on sales offices, see Chapter 6, Stage a Successful Sales Environment. For details on model homes, see Chapter 7, Merchandise Your Model Home.)

Checklist 5.5 Excellent Exteriors

___ Provide wheelchair-accessible entrances with no steps or raised thresholds.

___ Clearly identify the sales office, inventory homes, and model homes with signs.

___ Contract with a landscape architect (for larger projects) or professional landscaper early in the build process to install attractive, seasonal landscaping and maintain it well.

___ Keep exteriors of the structures well maintained with fresh paint, unweathered wood, and clean stucco, brick, siding, and the like.

___ Clearly post open hours.

___ Post a phone number and web address for customers to use to get more information during off-hours. Stock a supply of your brochures outside during off-hours only.

___ Bring brochures and collateral inside during open hours so as not to discourage buyers from entering.

___ Add warm, inviting touches to the front porch such as benches, settees, rocking chairs, or Adirondack chairs.

___ Place colorful pots and window boxes of flowers and plants on the porch as the season allows.

___ Provide activities for children, such as a large chalk board, white board or video game to encourage the kids' votes and occupy them while parents browse the model.

___ If you are building in a large community, work with the developer to locate benches or other seating in shaded areas throughout the community so prospects can rest periodically or enjoy the surroundings, particularly in a retirement community.

___ Check the amount of shade that is available for buyers meandering throughout a large community. Erect awnings, pergolas and arbors with climbing vines, or plant additional trees if necessary.

___ Keep lawns and flower beds mowed and trimmed.

___ Check the condition of company vehicles bearing your logo or company name. Make sure they are relatively clean and in good repair.

___ Walk your construction sites and be aware of the image projected by your homes under construction.

___ Keep clutter and chaos to a minimum.

Off-site Messages

Prospects throughout the community will form impressions about your company each time they encounter your company name, website, Tweets, ads, logo—even your company vehicles. Do a careful check to make sure these elements project a first-class image.

Company Vehicles

Whether it's your salespeople driving prospects from site to site or your company executives driving a vehicle marked with your logo, Checklist 5.6 will ensure that your vehicles become a virtual model home on wheels.

Checklist 5.6 Company Vehicles

Salespersons' Vehicles

___ Choose vehicles that will convey an image of quality and success, and select a four-door vehicle for ease in getting in and out

___ Regularly wash and vacuum vehicles.

___ Don't allow smoking in the vehicles.

___ Keep the radio turned off.

___ Keep the vehicle free from trash, collaterals, and other clutter.

___ Don't put anything in the ash tray.

___ Don't hang anything on the rearview mirror.

___ Ensure the vehicle has plenty of gas ahead of time.

___ Wear seatbelts and ask your passengers to do the same.

___ Drive safely and courteously.

___ Don't apply bumper stickers or decals, particularly of a political or divisive nature.

Other Company Vehicles

Although construction management and other personnel may not drive prospects in their vehicles, they will still be visible on-site or in the community at large. Consider the following list of items to be sure these vehicles convey the right image:

___ Display company name and logo prominently. (Magnetic signs are more affordable than applying directly on vehicle.)

___ Keep vehicles clean and in good repair.

___ Impress upon all personnel the importance of courteous driving while in company vehicles.

___ Choose a vehicle that will convey an image of sturdiness, reliability, and utility, such as a pick-up truck or sports utility vehicle.

___ Don't apply bumper stickers or decals, particularly of a political or divisive nature.

Creative Collaterals

The graphics, brochures, and ads representing your building firm or community reflect the quality of your firm and the homes you build. Listed in Checklist 5.7 are the elements they should have. For more on collateral materials, see Chapter 9, Create Collaterals with Impact.

Checklist 5.7 Creative Collaterals

Stationery

While a great deal of communication with your customers is online, there will still be instances when you will want to communicate with them in writing. Below are some tips to follow.

___ Print with same color as other graphics.

___ Clearly display name, logo, address, phone and fax numbers, and email and website addresses on letterhead.

___ Print note cards to match stationery for prospect follow-up and thank-yous to real estate agents.

___ Choose high-quality, textured paper.

___ Provide all salespeople and personnel with business cards that match stationery.

Brochures, Flyers, Floor Plans, Options List

___ Have printed materials professionally produced and printed—not photocopied.

___ List the benefits of your homes.

___ Select high-quality paper.

___ Include company logo, address, phone and fax numbers, and your email and website addresses.

___ Use the same color and style as other graphics for consistency.

___ Include a photograph or rendering of one of your homes.

Advertising and Direct Mail

See Checklist 5.8 for tips on how to make your ads and direct pieces shine. For more on advertising, see Chapter 8, Develop Your Advertising Program.

Checklist 5.8 Advertising and Direct Mail

___ Have ads professionally produced by a graphic designer or an advertising agency.

___ Keep style, message, and quality consistent with other graphics.

___ Clearly display logo.

___ Graphics and headline should be attention-grabbing.

___ Convey the benefits of your homes through your ads.

___ Keep headlines concise but ensure they deliver a strong message.

___ Don't overwhelm the space with too much copy; include generous amounts of white space.

___ Include a photo or illustration of one of your homes.

___ Include location(s) or directions, phone number, email, and website address.

Now that you have reviewed the quick checklists in this chapter, you need to look at some of these items in more detail, starting with sales offices.

STAGE A SUCCESSFUL SALES ENVIRONMENT

6

When prospects arrive to inquire about your homes, what kind of setting will you stage to meet them? Sales environments have changed drastically over the last ten years. Builders across the country are using technology to showcase the floor plans, elevations, features, and designs they previously mounted on every available inch of wall space. This chapter will help you meet today's prospect on their technological level, assess your image and credibility when prospects step into your sales environment.

Choose a Setting

Many factors, including your annual sales volume and whether you sell in a single community or on scattered sites, determine where you will meet your prospects. Listed below are some options you can consider. Determine the one that best applies to you or that you think is most feasible for you at this time.

You can make presentations in a:

- **Sales office.** This room or facility is a designated area, often converted from a garage. It is where the builder or sales representative meets with prospects and makes sales presentations. Sometimes the sales office is in a portable building or trailer. Figure 6.1 shows a typical floor plan for a garage converted into a sales office.
- **Model or inventory home.** You can greet customers inside the model home or inventory home, without doing any conversion to build a separate sales office facility. If done right, a model home can be an effective sales tool for the small-volume builder. Figure 6.2 shows a floor plan for this type of sales office.
- **Administrative office.** For a small-volume builder, the administrative office may mean his or her own home or leased office space.
- **On the go.** With today's tablet technology, a small-volume home builder can meet prospects in a local coffee shop or other convenient public spot.

Consider the attributes your sales office needs to convey— a message of excellence and quality— as well as the type of setting that will make visitors comfortable. Checklist 6.1 provides the elements of a successful sales environment for all three types of office arrangements.

Figure 6.1 Two-Car Garage Conversion

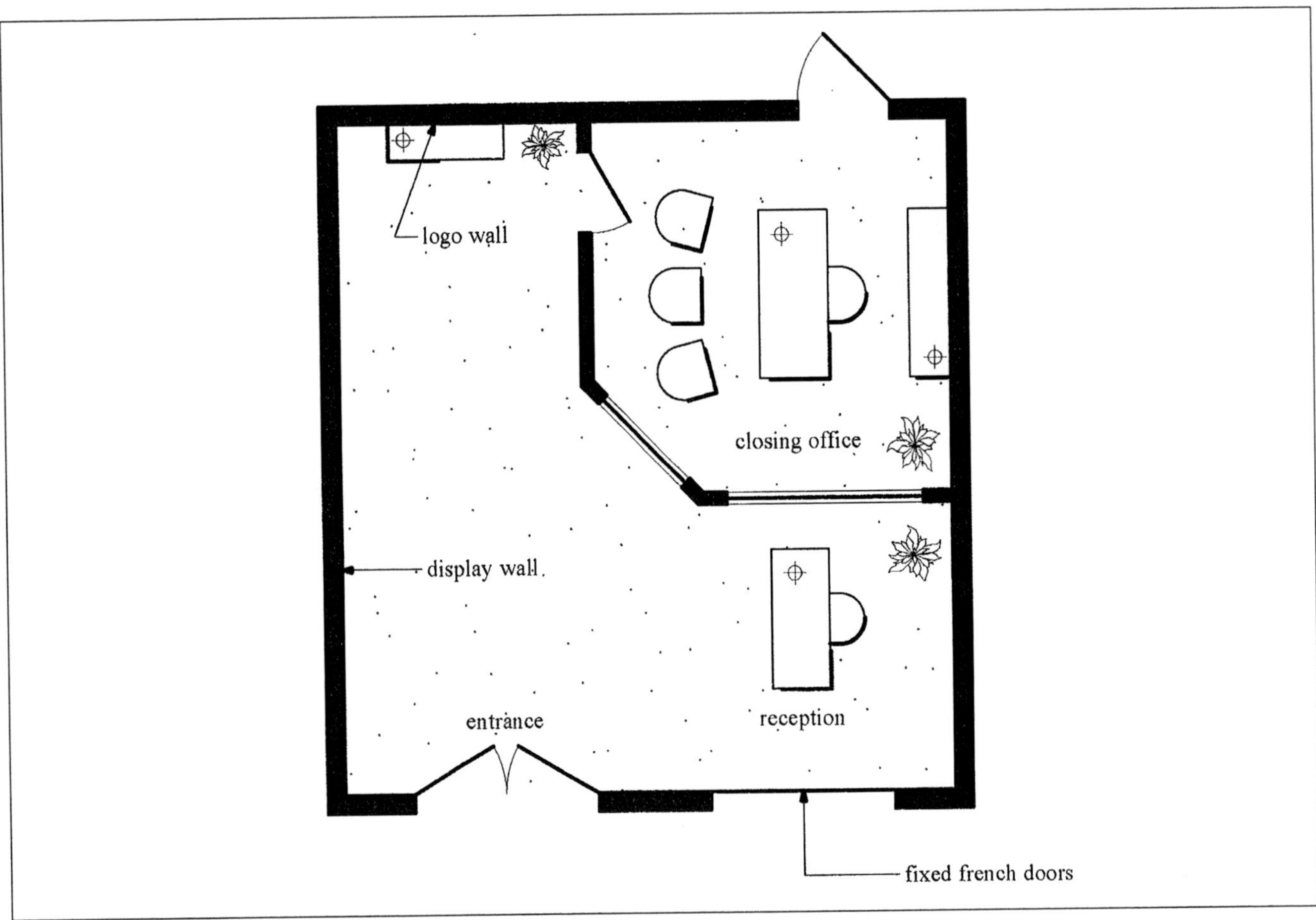

Reprinted with permission from Models, Inc., Chantilly, Virginia and Carlyn and Company Interior Design, Great Falls, Virginia.

Figure 6.2 Sales Office in Den or Study

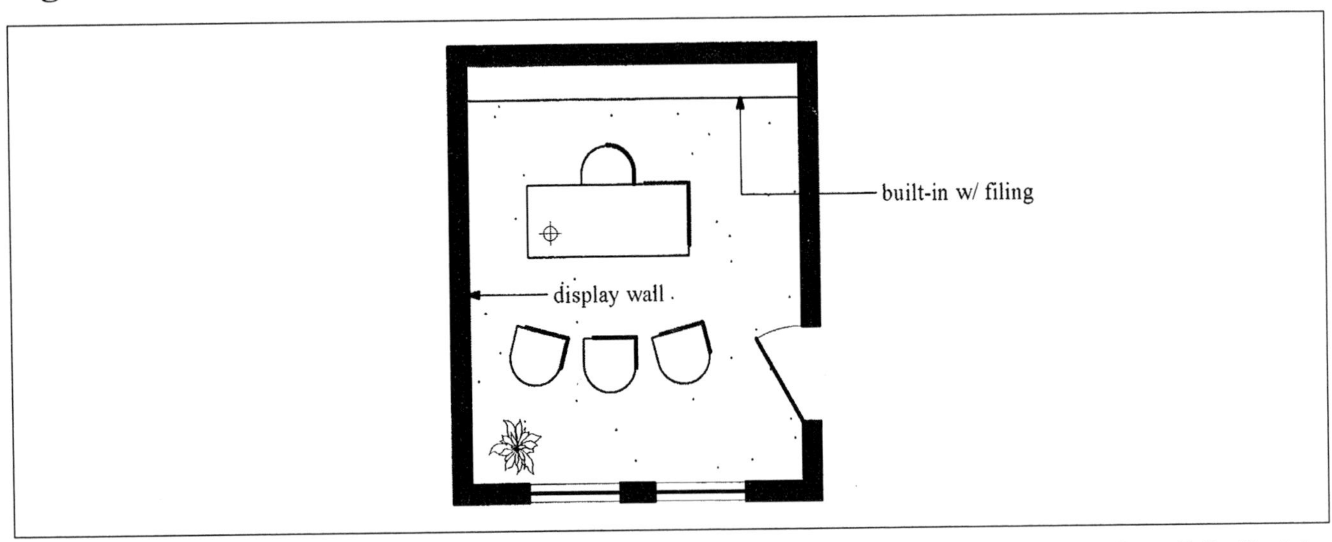

Reprinted with permission from Models, Inc., Chantilly, Virginia and Carlyn and Company Interior Design, Great Falls, Virginia.

Checklist 6.1 Successful Sales Environment

A Designated Sales Office

___ Design a warm home-like environment with wood floors and high-end detail.

___ Provide plenty of ambient and task lighting.

___ Design an open floor plan with plenty of windows and glass.

___ Decorate with current trends and styles and with the same theme as the model or inventory homes.

___ Erect exhibits and lifestyle photography on the walls.

___ Show photos of the local community, such as main street shops, a popular downtown square, etc.

___ Include a reception area, and furnish it with the following items:
- A registration table with brochures
- Two or three chairs or a sofa for casual seating
- One or two computer monitors or mounted tablets so that prospects can browse your uploaded features and visuals.

___ Accessories and plants to warm the area

___ Provide a private closing area, typically 10 ft. × 10 ft., in which to discuss pricing and financing issues. Furnish it with the following items:
- An executive desk and chair
- At least two guest chairs
- For a family market, include a basket of toys on one corner to occupy children.
- Space for computer, files, fax, and storage

___ Keep the office clean and well maintained. Care for plants, dust furniture, remove mud from floor, and vacuum carpet frequently.

___ Keep sales representatives' offices neat, clean, and free of clutter.

___ Provide a handicapped-accessible restroom. Keep it clean and functional and stocked with toilet tissue, paper towels, and hand soap.

___ Provide a water cooler or water bottles.

___ Stock refreshments such as coffee, soda, trail mix and fruit

Sales Office in a Model or Inventory Home

___ Designate a comfortable area in the home to meet customers, such as the den or office, or even a dining room. Or set up an office in a secondary bedroom.

___ Create a private space to discuss pricing and financing issues, such as a den or office with a door.

___ Furnish the space with a desk and three to four chairs.

___ Designate a downstairs bathroom for public use, and make sure it is spotless, well stocked, and functioning.

___ Limit refreshments to water or lemonade. Avoid dark-colored sodas or messy snacks.

___ Be vigilant about cleanliness and maintenance; monitor the area continuously for such items as discarded papers, cups, or crumbs left behind.

___ Set up lifestyle and community exhibits, as well as floor plans and elevations on the dining room or office wall or even in a visible hallway.

Sales Office at Home or an Administrative Office

___ Designate an area in which to meet customers, such as a conference room or comfortable sitting area with coffee table, couch, and chairs.

___ Use furnishings and accessories that reflect the quality of your work and demonstrate current trends. Hire a designer, if possible.

___ Make sure your restrooms are clean, well-stocked, and available for customer use. If the office is in your home, designate one restroom near the meeting area for customer use.

___ If the office is in your home, provide a separate outside entry if possible.

___ Use your iPad or other tablet to demonstrate floor plans, elevations, and features.

The Exhibits

In any of the three sales environments, you can use exhibits to demonstrate your qualifications and to build customer confidence. Traditionally these have been wall-hung displays, but with today's technology, you can also demonstrate these items on your iPad or other tablet. Make sure all exhibits are designed with the same materials and graphics, typeface, and color for consistency.

Whether your exhibits are digital, physically hung on the wall, or a combination of both, Checklist 6.2 lists some important exhibit elements and what they should contain.

Checklist 6.2 Sales Office Exhibits

Listed below are some exhibit elements for any selling environment and what they should contain:

Builder's story

- Establishes the credibility of the builder
- Details the history of the firm
- Introduces the firm's principal or principals
- Explains the principal's experience in the industry
- Explains the company mission statement or philosophy

Area locator map

- Shows location of homes or community in relation to surrounding area
- Reinforces the benefits of the location.
- Shows elements of interest, such as schools, parks, shopping, airport, business district, and industrial areas.

Lifestyle photographs

- Hire a professional photographer to take them.
- Show residents with the same demographic profile as your target market. Remember to adhere to Equal Housing Opportunity requirements for showing people in ads. See Complying with Fair Housing Laws in Chapter 8 for details.
- Show familiar shots of favorite community locations
- Show residents at recreation or leisure.
- Use some of the same photographs on your website, in ads and/or brochures.

Product photographs or renderings

- Make sure they reflect the quality of the homes.
- Use some of the same photographs on your website, in ads and/or brochures.
- Have a professional photographer or artist create them.

Testimonials

- Frame and mount letters, thank-you notes, or other correspondence from satisfied customers, or include them in a portfolio or scrapbook.
- Display any published articles about your company or community that have appeared in the local press in a framed format.
- With their permission, post on your social media sites, along with their photos.

Considerations for Special Markets

If you serve a distinctly targeted consumer, you will want to provide special accommodations to meet their needs. Checklist 6.3 provides special considerations for sales offices targeted to the senior market and the young family market.

Checklist 6.3 Sales Offices for Special Markets

Sales Offices for Senior Market

___ Erect exhibits and displays at heights for an audience that averages 5½ ft. tall.

___ Ensure lettering on exhibits is large enough so it's easy to read.

___ Depict people in lifestyle photographs approximately 10 to 15 years younger than actual target market.

___ Furnish sales environment with comfortable couches and chairs, but make sure the seating allows for ease in sitting down and getting up.

___ Keep paths and walkways clear, well-marked, level, and well lighted.

___ Make all areas wheelchair-accessible, as the American Disabilities Act requires.

___ Avoid yellow in furnishings and decoration because as eyes age, they tend to perceive colors with a more sallow cast.

Sales Offices for Young Families

___ Provide a play area or tot lot for children. It can be as simple as a basket of toys in a corner, as elaborate as a mini-playground, or as high-tech as a computer or video room.

___ Give collateral materials to children to keep, such as coloring books with the builder's logo, web address, and phone number.

___ Use lifestyle photographs to depict families and children.

___ Keep healthy snacks on hand that will appeal to children, such as mini-water bottles and fruit wraps.

___ Frame and display artwork by local school children who have visited your site or who have moved into the community.

___ Provide a changing facility, include a changing table in the restroom, or equip a model with a changing table in a merchandised nursery.

Consider High-Tech Options

Whether your target market is young professionals, experienced executives, or move-up buyers, chance are they will be familiar with the latest mobile apps, social media and tablet technology. And don't discount the boomers—they use the Internet as their primary means of comparison shopping for major purchases.[12] When these buyers arrive in your sales environment, their confidence level will increase when they see that you are a builder who uses the latest technology.

Figure 6.3 All Digital Sales Office

The I-BAR concept from The New Home Company is being employed at Lambert Ranch's sales office in The Grove, Irvine, California. iPads are clustered around a spacious table which accommodates several prospects at once. Customers can explore the builder's app and self-search for information and features. This has enabled The New Home Company to eliminate paper brochures and replace them with a digital brochure which can be downloaded into customers' smart phones. *Used with permission from The New Home Company.*

The All-Digital Sales Office

Consider eliminating topo tables and wall-mounted floor plans and elevations in favor of digitizing the information and showing it on a flat screen TV or computer monitor. (One method of powering this is via an Apple TV system connected to iPad tablet.) The advantages are as follows:

- Salesperson can take the tablet into the model as he or she conducts the model presentation.
- Customer can conduct searches and get more information on products, features, upgrades and brands included in the home.
- A cluster of iPads around a central table will become a gathering spot for prospects and provide a sense of community.

VIRTUAL MODELS

This technology allows customers to "tour" a three-dimensional furnished model. Listed below are some marketing advantages of using virtual models in your sales environment:

- Shows your homes in a life-like setting with furnishings, decor, and accessories. Allows the customer to view every room from different locations and angles.
- May be ideal for a small-volume builder with no models or inventory homes or for selling from a trailer site prior to construction or a meeting in a local coffee shop.
- Can be accessed online.
- Can be emailed to prospect

CONTINUOUS LOOP VIDEOS

Continuous loop videos can entertain prospects while a salesperson is busy with other clients by

- Showing clips of model interiors and surrounding community on a flat screen TV in the sales office or model home, and
- Supplementing your lifestyle photography, and projecting the style and quality of your homes.
- Now that you've staged the best sales environment possible, will your prospects move on to a fully furnished model to view your finished product? Or will you use another merchandising option? Read Chapter 7 to help you decide which on-site merchandising strategy is best for you.

MERCHANDISE YOUR MODEL HOME

7

One of the most important marketing decisions a builder can make is whether or not to build a model to help sell homes and, if so, how to present it in the best possible light. Sometimes a builder may decide to furnish and decorate an inventory home. Merchandising can sometimes determine the difference between a highly successful model and an ordinary one. Marketing experts agree that builders sell more homes when customers can see and touch the product and imagine themselves living there. A model home done right can be a powerful tool for creating an emotional response from your prospects. Even a small-volume builder can implement creative merchandising within a tight marketing budget.

When to Build a Model

Use the following criteria to determine whether you should build a model home or merchandise an existing inventory home. If any of the following situations apply, a model might work well for you:

- Your competitors in the same local market and price range are showing model homes.
- You are a new player in your marketplace.
- The unique selling points of your homes include their architectural features, upscale amenities, or dramatic finishes that are difficult to visualize before construction.
- Your budget allows for approximately 1.0% to 1.5% of your annual gross sales for merchandising and model home maintenance.

Model Home Exteriors

For a checklist to ensure that the outside of your model has strong curb appeal refer to Checklist 5.4, Excellent Exteriors.

Model Home Interiors

There are tasks that the builder or on-site salesperson should perform each morning before opening the model or inventory home to the public. See Checklist 7.1 to see what should be accomplished.

Checklist 7.1 Daily Model Home Tasks

___ Turn on all lights, lamps, HVAC system, music system.

___ Turn all ceiling fans to a low setting.

___ Pick up any litter, cups, trash, or crumbs.

___ Refresh the bathrooms

___ Empty wastebaskets and replace liners.

___ Restock supplies, such as paper towels, toilet tissue, and liquid soap.

___ Remove water spots from sink, fixtures, and mirror.

___ Check bathroom linens to see that they are hanging straight or provide disposable towels.

___ Straighten any collateral material that may be displayed in the home.

___ Clean up dust, insects, or tracked-in mud.

___ Employ a cleaning service to do a thorough cleaning at least once a week, ensuring that floors are vacuumed, furniture is dusted, and accessories are gleaming.

___ Shake dirt off the floor mat at the entry.

___ Set thermostat to a comfortable temperature.

___ Straighten and fluff pillows on couches, chairs, and beds.

___ In bedrooms, make sure bed linens are straight and that mattresses and box springs are covered.

___ Remove any leftover food from the refrigerator.

___ Turn on any fountains or water elements that can be seen or heard from inside.

Understand Merchandising

Merchandising is the art of equipping a builder's model or speculative home with carefully selected items so the home:

- Appeals to the intended target market and helps them visualize living there
- Enhances the architectural features of the home, such as staircases, windows, crown moldings, ceilings and windows
- Draws attention to views
- Capitalizes on the available space in all rooms of the home
- Downplays or disguises problem areas, such as awkward traffic flow or unattractive sight lines

Don't confuse merchandising with interior design. Interior design uses furnishings and decor to make a home attractive rather than focusing on selling the space or marketing to a particular buyer profile.

What to Use in Your Model

There are a variety of furnishings, built-ins and accessories you can use to merchandise your model or inventory home. Not all of these need to be or can be used in every model. Your merchandiser will know which ones are appropriate for each floor plan and buyer profile. Checklist 7.2 is a list of many of the possibilities that you can use in merchandising your model home.

Checklist 7.2 Model Home Furnishings & Accessories

FURNISHINGS

Living Rooms and Family Rooms

- Sofas, love seats, sectionals
- Chairs, ottomans
- Cocktail tables
- End tables
- Entertainment centers
- Faux props, such as flatscreen TVs and computers

Bedrooms

- Beds, headboards
- Night stands
- Dressers, chests
- Chairs, settees
- Desks
- Armoires
- Exercise equipment in alcove

Dining Rooms and Breakfast Nooks

- Table
- Chairs
- Buffets, hutches, sideboards

Decorator Items

- Window treatments, such as valances and curtains, shutters
- Paint (accent colors, faux treatments, walls, trims and stripe details)
- Stenciling and graphics
- Carpet, area rugs
- Flooring
- Mosaic tile work in kitchen and baths
- Bathroom fixtures (toilet, bidets, vanities)
- Baths, showers, saunas
- Hardware (Many manufacturers offer matching suites of hardware for bathrooms and kitchens)
- Light fixtures

Dens and Offices

- Desks
- Chairs
- Faux computer monitors
- Bookshelves, especially built-in
- Framed "family" photos (Use photos of people closely matching your buyer profile)

Patios, Decks

- Outdoor dining table and chairs
- Bistro table and chairs
- Umbrellas
- Picnic tables
- Chaise lounges
- Benches
- Wicker furniture, Adirondack chairs
- Barbeque grills, especially built-in
- Outdoor kitchens, cooktops, sinks, refrigerator, wine coolers
- Fireplaces/firepits
- Flat screen TV

ACCESSORIES

- Table and floor lamps
- Plants and flower arrangements
- Framed family photos (depicting people in target market)
- Framed art
- Candle sticks
- Mirrors
- Books, reading glasses
- Pillows

- Blankets, afghans, quilts thrown over backs of chairs and sofas
- Bedding
- Bathroom linens
- Kitchen canisters
- Place settings and table linens on dining tables and at eat-in bar
- Cookbooks
- Centerpieces for tables
- Coffee tray
- Wine glasses and wine bottle
- Soaps, bath accessories
- Hangers, hatboxes, shopping bags from popular local stores (for closets)
- Sculptures, "antiques"
- Lifestyle items such as lunch boxes for children and golf clubs for adults

BUILT-INS

The architectural elements and built-ins listed below can be strong additions if you plan them with the merchandiser in the product development stage.

- Plant shelves
- Book shelves
- Media centers (for placement of television and other electronic equipment)
- Desks
- Art niches
- Technology niches/recharging stations
- "Stop and drop" niches
- Crown molding treatments
- Chair rails
- Beadboard paneling
- Wood ceiling treatments
- Built-in barbeque grills

Make Vignettes Work for You

Vignetting is the art of staging the rooms in an unfurnished builder's model or speculative home by adding decorator touches and accessories that will appeal to the intended target market and will accent the architectural features without fully furnishing them. Listed below are some tips to keep in mind when creating vignettes:

- They can successfully appeal to the target market of a small-volume builder with a modest budget. Vignetting demonstrates the space you cannot afford to merchandise.
- Patterned fabrics and bold colors will make a stronger statement in vignettes than in a fully furnished model.
- Special details such as hardwood floors, special ceiling treatments, built-ins, decorative tiles, and architectural elements can become the focus of attention in vignettes.

Figure 7.1 Lifestyle Merchandising

Whether you fully merchandise your model home or use vignettes, careful placement of lifestyle accessories will appeal to your targeted buyer. Black cabinets with white Silestone counters and popular gray glass tiles make a dramatic statement as well. Model from Jimmy Jacobs Custom Homes at Caballo Ranch, Cedar Park, Tex. Merchandising by Mary DeWalt Design Group. Photographer: Rachel Kay, Applebox Imaging. *Used with permission from Mary DeWalt, MIRM, Mary DeWalt Design Group.*

See Checklist 7.3 to see what you might use in your vignettes.

Checklist 7.3 What to Use in Vignettes

- Plants with lighting behind them
- Framed art
- Mirrors
- Books (on built-in shelves)
- Bathroom linens
- Soaps, bath accessories
- Kitchen canisters, utensils, cookbooks, children's lunch boxes
- Dishes, table linens (at eat-in kitchen bar)
- Hangers, hatboxes, shopping bags from popular local stores (for closets)
- Fireplace accessories
- Mantel decor: candles, sculpture, etc.
- Rolled towels in a wicker basket in laundry room
- Entry table in foyer

Built-Ins

- Plant shelves
- Book shelves
- Technology centers (for placement of flatscreen TV, laptop, etc.)
- Desks
- Art niches
- Entertainment centers (for placement of flatscreen TV and audio controls)
- Crown molding treatments
- Decorative wall treatments
- Ceiling treatments
- Chair rails or beadboard

More Vignetting Strategies

- Painted murals in a children's room
- *Trompe l'oeil* (fool the eye) wall murals to depict how furniture will look in a room, such as a painted headboard on a bedroom wall or a painted window with flower box to open up a small space. You can have the murals painted on strippable wallpaper for easy removal.
- An easel with a framed color rendering of the room fully furnished

Partner with Retailers, Designers, Suppliers

Partnering with a local retailer or designer to furnish and decorate your model or inventory home can help reduce merchandising costs. Feature promotional signage or brochures for those cooperating suppliers and retailers in exchange for their assistance. The following are some ideas that you can explore for such a partnership. See if they will:

- Donate items for your model
- Lend items for your model
- Provide discounted items for your model
- Provide training services or videos for your salespeople
- Provide consumer information brochures or videos on certain brand-name products to create brand awareness, educate buyers on product use, and impart an image of high quality

Below are some ideas of products you can use in a partnership arrangement in your model or inventory home.

- Closet organizing systems
- Bedding and linens
- Accessories
- Window treatments
- Mirrors
- Framed art
- Plants, flower arrangements
- Paint
- Furnishings
- Built-ins
- Fireplaces
- Heating and air conditioning systems
- Brand-name windows
- Brand-name solid-surface countertops
- Brand-name appliances
- Brand-name bathroom and kitchen fixtures and faucets
- Hot tubs/spas

Cost-Cutting Merchandising Ideas

Below are some tips on how to keep the budget in check when merchandising.

- In an unfurnished model, provide templates of furniture with your floor plans so buyers can create their own room arrangements and decide how their own furniture will fit in your model.
- In an unfurnished model, provide suggested floor plan furniture layouts in your model. Have a professional interior space designer create these furniture layouts.
- Use your furnishings for more than one model, but update them with new accessories, such as window treatments, comforters/duvets, towels in the bathrooms, and pillows on a sofa.
- Replace furnishings after three to five years to keep everything fresh. Replace items immediately if they show wear.
- Lease furnishings and accessories from a furniture company. This can be economical if you will use the items for less than 2 years.
- Sell furniture back to supplier or to customers after model closes.

- Sell the furniture and décor with the model
- Use a combination of merchandising and vignetting. Fully furnish only those rooms that are most important to buyers—entry, kitchen, family room, and master suite; vignette the bathrooms and secondary bedrooms.
- Use paint instead of furnishings to create focal points in certain rooms, such as a headboard in a bedroom or a dramatic stripe detail on a wall. (See Figure 7.2)
- Use flea market or consignment store "finds" to decorate, such as a repainted old door for a headboard (for the right market).
- Use simple valances instead of full draperies.
- Use bound carpets (from your carpet supplier) instead of expensive area rugs.
- Use shelving and built-ins instead of furnishings to create focal points, such as entertainment centers in family rooms or focal points above beds.
- Merchandise the first floor only, if the model has a master bedroom downstairs.

Figure 7.2 Creating Drama Cost-Effectively

As this paint stripe detail shows at City Centre in Kitchener, Ontario, a dramatic effect can be achieved at a very low cost. Builder is Andrin Limited and merchandiser is Possibilities for Design. *Used with permission from Doris Pearlman, MIRM, Possibilities for Design.*

Budget-Cutting Tip

If you have a previous customer who has decorated their new home beautifully and in keeping with your targeted audience, feature an open house in their home in exchange for free upgrades or for a generous gift certificate. Also offer to provide housecleaning before the opening.

Design Centers

Larger builders or builders within a planned community may offer their buyers a design center in which to choose their options, finishes and colors, allowing customer the ability to customize. For the smaller builder or the builder on a limited budget, there are other ways to accomplish this. Consider the following choices:

- Actual product samples mounted on boards
- If your sales environment is a model or inventory home, display the designer selections in a secondary bedroom.
- Visual images on a iPad or tablet

Whatever the format, see the items in Checklist 7.4 for the items that can be displayed.

Checklist 7.4 Design Center Choices

- Exterior choices (such as brick, siding, stone or cultured stone, or choices of stucco color)
- Flooring choices
- Kitchen countertop samples
- Appliance packages
- Bathroom vanity samples
- Carpet samples
- Paint colors (exterior base and trim; interior wall and trim)
- Fireplace mantels and surrounds
- Interior door styles
- Cabinetry and hardware choices
- Stair rail choices
- Lighting fixtures
- Kitchen faucets
- Bathroom faucets and fixtures
- Landscaping packages
- Outdoor kitchens, built-in BBQs
- Options such as security systems, sound systems, upgraded wiring, etc.

Ensuring a Successful Model Home

Checklist 7.5 includes safeguards that will help ensure a smooth model home experience, from installation to the customer presentation.

Checklist 7.5 Ensuring a Successful Model Home

___ Contract with your merchandiser in the planning stages of your model to ensure that floor plans have good traffic flow, attractive sight lines, and most efficient use of space.

___ Continue to include your merchandiser in your marketing meetings so he or she can make the merchandising effort consistent with your overall marketing program.

___ Ensure you are working with a qualified model home merchandiser by checking with the NAHB Institute of Residential Marketing (IRM) to find professionals who are IRM-certified. See the NAHB Directory of Professionals with Home Building Designations at http://tinyurl.com/knb9edq.

___ Target your model home toward a specific buyer profile, even a specific fictional family.

___ Use a consistent style or theme throughout the home in furniture and color.

___ Use furnishings and decor that are consistent with the style of your home. (For example, don't put contemporary furniture in a home with Colonial-style architecture).

___ Create a memorable focal point in main rooms, such as a fireplace in the family room or mosaic tile behind cooktop in a kitchen.

___ Don't cover architecturally beautiful windows or views with heavy or overabundant window treatments.

___ Scale furniture to the model's size; for instance, don't overcrowd a small floor plan with large pieces.

___ Use color appropriate to the model's size; for example, use warm colors to fill up a large room and, in smaller plans use light colors on walls, floors, sofas, and beds.

___ Accentuate architectural details with artwork or window treatments.

___ Place furniture so it creates pleasing sight lines and smooth traffic flow.

___ Keep cabinets and closets free of clutter. Prospects will look inside!

___ Merchandise closets with shopping bags, fluffy robes, and suitcases.

___ Use durable flooring to ensure that it holds up to model home traffic.

___ Provide ample lighting, a minimum of 100-watt bulbs in central fixtures throughout (unless the fixtures specify otherwise) and have all lights on when you show a model.

___ Use extra lamps to create brightness and ambiance.

___ Install in your model any upgrade lighting packages available.

___ Use accent lighting, such as can lights, track lighting, and recessed lighting to enhance architectural features.

Merchandise for Your Target Market

For merchandising tips for multicultural buyers, see Chapter 12, Appeal to the Growing Multicultural Markets. For merchandising tips for all other markets, consult the merchandising sections in Chapter 3, Develop Your Marketing Strategy.

Choose a Model Merchandiser

Use a merchandiser with professional credentials who has new home experience and a background in marketing, rather than a friend who may be good at decorating. Beyond just a good decorator, a model home merchandiser has the expertise to review and critique floor plans to ensure furniture placement, sight line opportunities, traffic flow, and other lifestyle requirements. A model merchandiser who wants to work with you should provide a written proposal and a professional presentation, complete with color boards, floor plans, elevations, furniture packages, and a portfolio.

When you select a professional to merchandise your model or inventory home, make certain he or she has the skills, experience, and capabilities in Checklist 7.6.

Checklist 7.6 Choosing a Model Merchandiser

Make sure your model merchandiser offers the following qualifications:

___ Experience with newly-built homes

___ Knowledge of target marketing

___ Credentials from any or all of the following professional associations:

- Member, Institute of Residential Marketing (MIRM), nahb.org
- Society of Interior Design (ASID), asid.org
- International Interior Design Association (IIDA), iida.org
- Association of Interior Design Professionals (AIDP), aidponline.com

___ Can give referrals of satisfied builder clients

___ Has experience in or knowledge of your market area

___ Understands current lifestyle trends

Make sure the merchandiser will:

___ Walk through the model home at the drywall stage to check for any obvious floor plan flaws or traffic flow problems and recommend changes as needed.

___ Conduct training sessions with your sales staff prior to model opening to show the best way to demonstrate the merchandised space and sell the architectural features. (For more details on this, see Checklist 7.7)

___ Contact builder regularly to ensure there are no items that need repair or replacing.

The Merchandiser and Your Sales Staff

A model home provides a salesperson with the opportunity to create conversation and form a bond with a potential buyer. Before your model opens, your model merchandiser should meet with your sales staff to conduct a training session, to orient them to the benefits and features of the model, and to explain the strategies for merchandising the model. Use Checklist 7.7 as a guide for your model merchandiser and sales staff to work together.

Checklist 7.7 The Merchandiser and Your Sales Staff

In order for your sales staff to benefit the most from the expertise of your model merchandiser, ask your merchandiser to do the following:

___ Walk the model with the sales staff

___ Conduct an orientation on the buyer profile. (For example, a home is merchandised for a move-up family of four. The parents are between 25 and 39 years of age. They have one infant girl and one school-aged boy.)

___ Offer an alternative presentation for use with other types of buyers. (For example, tell a dual-income couple with no kids how they can use secondary bedrooms for his-and-her offices or exercise rooms.)

___ Demonstrate the best places to stand, move, and sit in each room to maximize the merchandising.

___ Explain why particular furnishings and accessories were chosen and how they appeal to the targeted buyer profile.

___ Where applicable, always indicate where upscale materials and products are standard, stating, "it's included!"

___ Explain how the spaces can flex for more than one function, for example, an eat-in island can serve as a serving buffet for large parties or a sleepover window seat in a child's bedroom.

___ Point out how certain items highlight architectural features. For instance, a dramatic wall-hanging might draw attention to crown molding.

___ Show how window treatments and judicious furniture placement enhance views.

___ Review the standard items included in the base price and those that are optional designer features, such as stainless steel appliances or built-ins.

___ Discuss maintenance and care of certain furnishings and accessories, such as a leather sofa or hardwood floors.

___ Provide a floor plan, video, or photograph showing proper placement of all items so the sales staff can ensure cleaning crews replace them correctly after cleaning.

Current Trends in Merchandising

Below are some of the general trends that some of the nation's award-winning model homes are demonstrating.

- Fewer matched-set furniture suites in favor of an eclectic mix of styles to appear as if a family accumulated the items over time.
- A global feel with accessories that look as though the family has collected them through their travels.
- Clean, straight lines and a minimalistic look for a more contemporary feel.
- Less clutter, such as eliminating ivy on top of cabinets.
- Fewer knick-knacks for a more minimal look.
- Use of color to create memorability, such as an accent wall or a striped detail.
- Built-in "stop and drop" locations where home owners can drop their mail, keys, etc.
- Organized mud rooms with cubbies for each family member's boots, backpacks and coats.
- A "retail catalog" look in furnishings and accessories, such as found in popular catalogs Crate and Barrel, Pottery Barn, West Elm, and Restoration Hardware.
- Demonstrating environmental consciousness using materials such as recycled glass countertops, salvaged wood furniture and sustainable flooring, such as bamboo.
- Multi-room audio/video systems that will allow home owners to play music and video throughout the home via a central source, smart phone or tablet.
- Security and thermostat systems that are controllable by smart phone and offer text notifications, such as when children arrive home from school.
- One bedroom or a niche merchandised as an office complete with computer or flat-screen monitor. (You can use faux props.)

- A family, game, or bonus room demonstrated as a media room or home theater with surround sound, flat-screen TV, popcorn machine, and theater seats.
- A bedroom, loft, or sitting area off the owner's suite merchandised as an exercise area with yoga mat, exercise ball, weights, and/or an exercise bike.
- A secondary bedroom shown as a hobby room featuring crafts that are trendy or popular, such as beading, painting, or scrapbooking or canning in the kitchen.
- Cozy sitting rooms with a fireplace off the kitchen or a romantic master bath complete with candles, wine bottle and two wine glasses.
- A teen retreat with computer desks, study area, game tables, Twister game, mini refrigerator.
- Recycling bins in kitchen and laundry rooms.
- Closet organization systems, especially in master suite.
- Closets and linen cabinets accessorized with hat boxes, hangers, and shopping bags.
- Kitchen pantries accessorized with boxes of cereal, granola bars, and other staples.
- Computer niches, chalkboards, or erasable whiteboards in children's room
- Upgraded cabinet packages in kitchens, such as glass-front doors
- Lazy Susans for corner cabinets, pull-out spice storage and wine bottle and glass racks.
- Mosaic tile backsplashes or subway tile in kitchens.
- Butler's pantry between kitchen and formal dining room.
- Upgraded lighting packages, such as can lights, rheostats, and under-counter lighting in kitchen.
- Kitchen islands showing multifunctions: eating, rinsing vegetables, chopping station with built-in cutting board.
- Pendant lighting hanging over kitchen island
- Architecturally stunning oven hoods made of stainless steel or of wood matching the kitchen cabinetry
- Built-in refrigerators to match kitchen cabinetry
- Hardwood floors or stone tile such as slate or limestone.
- Iron light fixtures and patterned iron stair railings
- Create wine storage in the basement or a wine closet off the dining room.

After completing this chapter, you should be able to better assess whether or not you will build a model and, if you do, the right way to show it. But whether you build a fully furnished model or vignette an inventory home, your next task is to bring qualified prospects out to see it. Turn to the next chapter to learn how to get the word out through advertising.

DEVELOP YOUR ADVERTISING PROGRAM

8

You now have a well-conceived product, a super sales office, and even, perhaps, a marvelous model home. But all your carefully targeted marketing efforts will be wasted if you don't draw a steady stream of qualified traffic. How do you get the word out to the right prospects and motivate them to visit your site? With a carefully planned advertising campaign thematically tied to the rest of your marketing program, you can help bring buyers to your door.

Review Your Research

You should build your advertising program on the market information you gathered in Chapter 1. Then, 1) develop your advertising strategy, 2) select which media to use, and 3) schedule your media messages. To help you narrow your focus, review the following items:

- Primary and secondary (if applicable) target markets
- Price range of homes
- Goals (For example, sell as many homes as the current market leader in your price range by the end of this year.)
- Your position in the marketplace (For example, you will be the only builder in the area who provides affordable homes with luxurious finishes for move-up buyers.)
- Gross annual sales projection for the current year
- Percentage of projected annual sales you will allow for your advertising budget (A typical advertising budget for a small- to medium-volume builder, including print, electronic, and other media, ranges from 1.75% to 2.5%
- Challenges you must overcome with your advertising program (For example, you are entering as a new player in a well-established market, must overcome the prospects' negative perceptions of previous ownership, or simply need to increase traffic to your homes or community.)

Create Your Logo

The crux of your advertising program is your company logo. A graphic artist or an advertising agency can help you create a logo for your building firm. See Checklist 5.1, Logo Design, to review the qualities of an effective logo.

Determine Your Media Mix

The best vehicles for delivering your message will vary according to your targeted buyers, local market conditions, and costs, but you probably will want to employ a combination of media. Listed below are descriptions of the most common types used by builders.

Online

Google AdWords and Microsoft Ad Center are two examples of pay-per-click ad services. In this model you only pay when someone clicks on your ad. When you purchase the ad, you designate how many clicks you want to pay for and the ad will continue to appear until it gets that many clicks. The ad appears based on a predetermined list of key words, such as "Scottsdale new homes." If an interested person clicks on the ad, the user pays a fee.

Advantages

- The advertiser is only charged if someone clicks on their ad
- Reaches a large audience
- Results are measurable

Disadvantages

- Technical difficulties sometime occur, reducing the time your ad may be seen.
- Consumers can bypass your ad by not clicking on a banner or by disabling pop-ups.

Facebook

Begin by developing a Facebook page, then a series of ads. When a user clicks on your Facebook ad, they will be directed to a destination of your choosing, whether it's your Facebook page, your website, blog, etc. You are charged only for the number of impressions or clicks the ad receives. When you begin you set a daily or lifetime budget and will never be charged more than that amount. Naturally, the larger your budget, the more people the campaign will reach.

Advantages

- Offers targeting options to such as gender, age and education to reach the audience most relevant to your business.
- Promoting a contest offers people an easy, low-pressure way to contact you, bringing more people to your site

Disadvantages

- Facebook ad formats may not allot enough space for images or text.
- Larger companies can bid more per click for their ads, potentially making it harder for smaller companies.

Newspaper

Despite the growing reliance on the Internet, newspapers are still finding a place in the home builder's marketing mix, typically by appearing in the real estate section of their local newspaper or weekly real estate supplement. You may want to consider other sections where your competition may not be advertising, such as in the lifestyle, business or home and garden section, or even in the classified ad section.

Advantages

- Can be less expensive to place ads compared to other types of media.
- Can be the most cost-effective medium for builders.
- Can be inexpensive to produce, especially in black and white.
- Well-targeted because prospects in the market for new homes will be looking here for information.

Disadvantages

- Losing readership to online portals
- Likely to be cluttered with competitors' ads
- Have a short shelf-life, usually no more than a day

Magazine

Place ads in local community, business, entertainment, or lifestyle magazines or (for a large-volume builder) in regional issues of national magazines. Also consider new home guides which are usually distributed free.

Advantages

- Highly targeted; publications often have new homes sections and publishers usually provide a demographic profile of their readers.
- Usually have minimal competition from other builders.
- Project a high-quality, upscale image.
- Have a relatively long shelf-life.

Disadvantages

- Usually more expensive to place than newspaper advertising.
- More expensive to produce than newspaper advertising because they often involve four-color photography and production.

Radio

These spots run on your local radio stations and are typically 30 seconds long.

Advantages

- Can target specific buyer profiles based on demographics from the radio station.
- Have little day-to-day competition from other builders.
- Are especially good for supporting limited *promotions* or special events such as grand openings.

Disadvantages

- Can be difficult to sustain listeners' attention; changing stations during commercials is so easy.

Television

These commercials run on your local television or cable stations. They are also typically 30 seconds long.

Advantages

- Can target specific buyer profiles based on demographics of an audience for a particular time slot.
- Have little day-to-day competition from other builders.
- Can project an image of quality, if professionally produced.

- Presents the option of producing stills with an accompanying voice over in order to minimize production costs; must be careful to come across professionally.
- Particularly effective as part of a total media campaign with radio, web, and print media
- May be affordable to place on cable television stations.

Disadvantages

- Too expensive for most small-volume builders to produce professionally; poor quality can be deadly.
- Usually expensive to place on local network affiliates.
- On cable channels, builders may face competition from resale homes because real estate brokers often use this medium.

Outdoor Advertising

Outdoor advertising appears on more than just billboards. It also includes ads found on bus shelters, bus and train stations, as well as inside and on the side of buses and trains themselves.

Advantages

- Can be relatively inexpensive to produce.
- Can be affordable to place, depending on your market area.
- Creates awareness and a strong presence in your community.
- Can direct prospects to your site.

Disadvantages

- Must be constantly maintained because of weather, wear, and vandalism.
- Delivers only limited information.
- Builder may have to commit to a minimum term that the ad will appear.

Create Effective Ads

Contrary to popular belief, the objective of your advertising is not to sell homes. It is to convince prospects to visit your site. Once they arrive, the on-site salesperson and the on-site merchandising must make and close the sale. Checklist 8.1 provides guidelines to ensure that your ads accomplish this objective.

Checklist 8.1 Effective Advertising

All Media

- Professionally produced and present a professional image.
- Include a map or directions.
- Brief and easy to read quickly. Not cluttered with too much copy, nor do they overwhelm the audience with too much information.
- Emphasize benefits, not just features.
- Include the Equal Housing Opportunity (EHO) logo or statement (see Comply with Fair Housing Laws later in this chapter for more information on complying with Fair Housing Laws).
- Color, design, typeface, or theme are consistent with other signs and graphics.
- Internet Ads

- Include website and email addresses.
- Eye-catching
- Offer contests, which entice readers to click on the ad for more information

Print Ads

- Include company name and logo.
- Include a phone number, email and website addresses for more information. Feature an eye-catching headline.
- Use photographs rather than renderings when possible.
- Include the Realtor® logo (if you are a member of your local Board of Realtors®) to show your willingness to cooperate with brokers.
- Include logos of any national warranty programs you may offer.
- Include logos of trade associations you belong to, such as NAHB.

Television and Radio Ads

- Use humor or emotional appeal to create memorability and prevent station-switching.
- Place on stations and in time slots with audience demographics in line with your target market.
- Run during times of high exposure, for example, during drive times.
- Make ads creative and attention-grabbing, without becoming irritating after repetition.

Outdoor Advertising

- Make the copy clear and large enough to be seen from a passing car.
- Use black or another bold color for the copy so it will contrast well with the background. Make graphics and headline immediately grab viewers' attention.
- Keep copy to a minimum; passing cars won't have time to digest too much information.
- Include directional arrow to your home(s), sales office, or site when possible.
- Include the price range of your homes and a telephone number.

Tricks of the Trade

Below are suggestions for making your ads stand out from the competition.

- Be creative with your use of white space. For example, leave a large margin around your ad or position your graphics in one comer.
- Make headlines in reverse type, that is, white type on black or other dark background.
- Make graphic elements or photographs break through the borders of your ad. For example, a sunburst pushing through the border.
- Run some or all of your ads in a part of the newspaper other than the real estate section, for example, the lifestyle or business section, depending on your target market.
- Use unusual or unconventional headlines when appropriate for the market. Examples that have been used are: "Rip Us Off," or "Throw Your Money Away! (Buy Someone Else's Home)."
- Run a promotional ad, that is, one that touts a special sale, grand opening, on-site promotion, or limited offer.
- Use bullet points instead of straight copy to highlight benefits for a cleaner look and more concise approach.
- Draw attention to special information, or new information not included on previous ads, through the use of boxes, banners, or cachets (the impression left by a rubber stamp or seal).

Figure 8.1 Ad for "The Station"

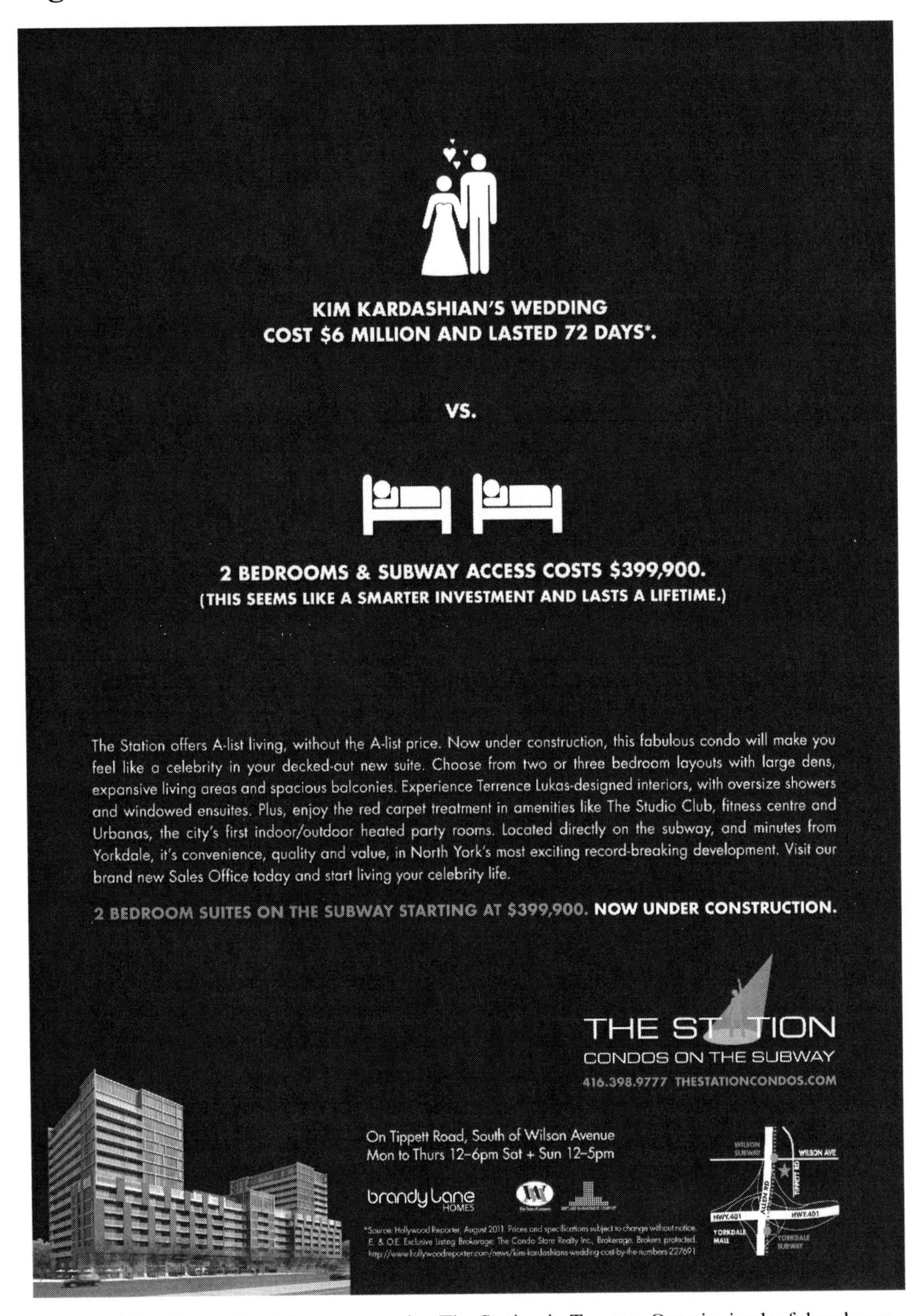

The brand for this condominium community, The Station in Toronto, Ontario, is playful and uses icons and *dingbats* as unique graphic elements to help it stand out from the competition. This black and white ad used a well-known celebrity reference to get the value message across and was a refreshing approach from the often tired, standard marketing. BrandyLane Homes won a Gold award for Best B&W Ad in the 2013 Nationals Awards sponsored by NAHB's NSMC. The advertising agency was LA, Inc. from Toronto. *Used courtesy of BrandyLane Homes.*

- Use humor in your ad, but use it appropriately. Don't be offensive or insulting.
- Use emotional appeal in your ad, for example, children playing baseball with their dad or a couple in a hammock.
- Use creative, unusual, or bold borders around your ads.
- Use a series of teaser ads. These ads entice the reader with bits of information and increase the details as the campaign progresses. This type of ad works best for print ads and billboards.
- Turn the ad at an unusual angle, or use an unconventional shape.
- "Moving" objects can protrude through your ad, or pop out as if they were three-dimensional.

Advertise to Your Target Market

For the advertising strategies that appeal to particular demographic groups, refer to Chapter 3, Develop Your Marketing Strategy. For advertising strategies targeted to multicultural buyers, see Chapter 12, Appeal to the Growing Multicultural Markets.

Figure 8.2 Ad Campaign for Eagle Construction

This ad campaign for Eagle Construction positions new construction homes as the preferred choice over resale homes. The simple layout with a lot of white space captures readers' attention. By showing obviously outdated styles, the ads demonstrate that a home built even ten years ago is outdated when it comes to energy efficiency, floorplan, and materials. This campaign won the Gold award for Best Print Campaign in the 2013 Nationals sponsored by NAHB's NSMC. Strategic brand manager was Phillip Glaeser and creative director, Maggie Miller. *Used courtesy of Eagle Construction.*

Working with an Advertising Agency

There are several ways to work with an agency:

- A flat fee for the creative work, plus a commission on media costs. Average range is 15 to 17.65% of what the media charges the client.
- On a per-project basis for time and materials, for example, copywriters and graphic artists charge for their fees, plus any photography expenses for your project.
- Monthly retainer. The agency charges a monthly fee for its continual services and consultation.

How to Choose an Advertising Agency

Before deciding on an advertising agency, narrow your search down to two or three, and then ask each the questions in Checklist 8.2, Choosing an Advertising Agency.

Checklist 8.2 Choosing an Advertising Agency

___ Does the firm have specific experience in advertising new homes or in real estate?

___ Does the firm have a proven track record, and is it able to provide clients' names?

___ Is the firm willing to work within your budget?

___ Will the firm conduct regular meetings with the builder and/or the marketing director?

___ Is the firm, a principal of the firm, an ad executive, or your project director a member of industry-related, professional organizations such as the:

- Local home builders association
- Local sales and marketing council
- National Association of Home Builders (NAHB)
- National Sales and Marketing Council (NSMC)
- Institute of Residential Marketing (IRM)
- National Association of Realtors® (NAR)

___ Do you understand and feel comfortable with the firm's billing procedures?

Weigh Frequency and Reach

In an ideal world an unlimited budget would allow you to place the largest ads any number of times and employ those media that reach the maximum number of people. But in reality, your budget will limit the strategy you choose. Therefore, when you decide on an advertising schedule (that is, how often to run which ads and which media to use), you must weigh the issue of *frequency* versus *reach*. Frequency is the number of times an advertising message is repeated. Reach is the number of prospects who will receive your single advertising message.

Listed below are some general guidelines to consider when you are developing your advertising schedule:

- Repeat an ad at least three times. One ad placement in a single medium is quickly forgotten.
- Remember that more, smaller ads are better than one big ad.
- Narrow the target audience and increase frequency rather than broaden the audience and reduce frequency.

Evaluate Your Advertising

There are three good ways to evaluate the effectiveness of your advertising: 1) Prospect Registration, 2) Exit Interviews, and 3) Tracking your "Calls to Action." Refer to Checklist 8.3, Evaluating Your Advertising Checklist, to learn the steps in executing each one of these.

Checklist 8.3 Evaluating Your Advertising

The following three ways can help you evaluate the effectiveness of your advertising program and help you decide if you are spending your ad dollars in the right places.

1. **Prospect Registration.** Using your CRM program, you should ask each visitor to your model or sales office how they learned about your homes, providing the following choices:

   ____ Internet Search (Which search engine?)

   ____ Social media site

   ____ Driving by/Directional signage

   ____ Billboards

   ____ Newspaper ad. Which paper?

   ____ Magazine ad. Which magazine?

   ____ Radio ad. Which program and station?

   ____ Television ad. Which channel?

   ____ A friend's referral. Who?

   ____ A Realtor's® referral. Who?

   ____ A mailing to your home

   ____ Other. Specify

2. **Exit Interviews.** A salesperson should ask all departing guests:

   ____ "How did you hear about us?"

   ____ "Were you referred by our website or other online portal?"

   ____ "Have you seen our current ads running in the newspaper, heard them on the radio, etc.?" Post tearsheets of your current ads in your sales office to stir prospects' memories.

3. **A Call to Action in your ads.** This strategy works for online, print or a direct mail campaign. Ask the prospect to do something, such as show the downloaded coupon on their smart phone, bring in a print ad or mention the radio station to receive a give-away or be eligible for a drawing. This way you can track how many of these people arrive at your sales office.

Comply with Fair Housing Laws

The United States Department of Housing and Urban Development issued regulations under the Fair Housing Amendment Act of 1988 that govern how housing may be advertised to the general public. By adhering to these guidelines you can avoid costly penalties and help to provide housing to residents on a fair basis.

Equal Housing Opportunity Logo, Statement, and Slogan

The Act states that you must include one of the following in your ads: the Equal Housing Opportunity (EHO) logo, the EHO statement, ***or*** the EHO slogan that is part of the logo, "Equal Housing Opportunity."

Figure 8.3 Equal Housing Opportunity Logo

Source: U.S. Department of Housing and Urban Development

The Equal Housing Opportunity Statement

We are pledged to the letter and spirit of U.S. policy for the achievement of Equal Housing Opportunity throughout the Nation. We encourage and support an affirmative advertising and marketing program in which there are no barriers to obtaining housing because of race, color, religion, sex, handicap, familial status, or national origin.

Source: U.S. Department of Housing and Urban Development, *Implementation of the Fair Housing Amendment Act of 1988*, 24CFR, Section 109.30, pp. 2–7.

Correct Use of Equal Housing Opportunity Logo

Most builders use the logo for print ads. If you use the logo rather than the statement, you must make sure the slogan is included in the logo and comply with these additional standards.

- If the ad is ½ page or larger, the logo must be 2 × 2 inches.
- If the ad is ⅛ page to under ½ page, the logo must be 1 × 1 inch.
- If the ad is 4-column inches to ⅛ page, the logo must be ½ × ½ inch.
- If the ad is less than 4-column inches, you must use the statement or slogan instead of the logo.
- If you use other logos in your ads (including your own logo, the Realtor® logo, your warranty company logo, and so on), the Equal Housing Opportunity logo must be at least as large as the largest of the other logos.[13]

How to Use Models in Your Ads

When you use people in your ads, either in photographs or drawings, you must adhere to the regulations set forth in the Fair Housing Act. See box, Models in Advertisements, for these requirements. Listed below are some strategies that builders use to help ensure that they comply with the Act:

- Emphasize product photos instead of using people in ads. The disadvantage of this strategy is that it may limit creativity and target marketing attempts.
- Feature an ad showing a group of people representing the same ethnic make-up as the community in which you are building. You may also be able to buy stock photography of this nature.
- Produce two or more ads featuring people representing the different ethnic backgrounds of the people in the community in which you are building. Rotate the ads on a weekly basis.

Models in Advertisements

Human models in photographs, drawings, or other graphic techniques may not be used to indicate exclusiveness because of race, color, religion, sex, handicap, familial status, or national origin. If models are used in display advertising campaigns, the models should be clearly definable as reasonably representing majority and minority groups in the metropolitan area, both sexes, and when appropriate, families with children. Models, if used, should portray persons in an equal setting and indicate to the general public that the housing is open to all and without regard to race, color, religion, sex, handicap, familial status, or national origin, and is not for the exclusive use of one such group.

Source: U.S. Department of Housing and Urban Development, *Implementation of the Fair Housing Amendment Act of 1988*, 24CFR, Section 109.30, pp. 2–7.

Tap into Advertising Resources

- Lifestyle Merchandising, Advertising, and Promotion Strategies (IRM III) is one of the four required courses for IRM certification. Along with merchandising and promotion, this NAHB course covers a variety of advertising media alternatives.
- To find advertising executives who are certified as a MIRM, refer to the NAHB Directory of Professionals with Home Building Designations at http://tinyurl.com/knb9edq. These professionals will have first-hand knowledge of advertising in the new home industry.

Once you have attracted prospects to your site, don't let them walk away empty-handed. Be sure to have collateral materials to give them. Often your collateral materials are produced together with your advertising. Turn to the next chapter to see how to create effective collaterals.

CREATE COLLATERALS WITH IMPACT 9

Your printed materials, or collaterals, can be some of the strongest tools in your marketing program. At the end of a long day shopping for homes, your prospects are likely to have toured dozens of models and gathered handfuls of brochures. Back at home, the most serious shoppers will sort through the collection of materials and try to compare builders. Therefore, your collateral materials must spark their memories, distinguish your homes from those of your competitors, and project a strong image. You don't want prospects to walk away from your sales office, model, or speculative home empty handed, and you want to make sure that what you provided has a powerful impact.

What to Include in Your Package

Your collateral package should include at least one, but can include a combination, of the following:

- Brochures, flyers, or folders
- Floor plans (sheets or inserts)
- Price, benefits, and features list
- Site map
- Community information
- Supplier and vendor information

Tech Tip

Today's mobile applications allow the flexibility of eliminating some or all paper brochures and collateral by downloading them into customers' smart phones. Then they can continue to shop at their own convenience when they arrive home.

Brochures, Flyers, and Folders

Your prospective buyers should get an informational handout in the sales office, model, or inventory home with facts, features, and benefits of your new homes. Brochures run the gamut from simple and inexpensive to elaborate and pricey. Tailor yours to your budget and your target market. Listed below are some commonly used formats:

- **Tri-fold leaflet.** This 8½ × 11 in. sheet folds in thirds. It is the most common type of brochure and one of the least expensive to produce. You can also produce a quarter-fold leaflet on 8 × 14 in. paper.
- **One-page flyer.** Usually a flat 8½ × 11 or 11 × 17 in. sheet, this flyer is the least expensive to produce. Its use is appropriate for a small-volume custom builder who may want to produce a piece for each speculative home in inventory or as a handout to distribute to real estate offices.
- **Folder with inserts.** The cover, usually 9 × 12 in., provides general information about the builder. The inserts are usually 8½ × 11 in. and relatively inexpensive to produce. They can include information that is subject to change. A folder is adaptable, which is an advantage if you do any of the following:

- Serve different target markets.
- Offer a variety of floor plans.
- Build in different communities.
- Build different product types.
- Introduce price increases or discounts periodically.

- **Multipage booklet or "coffee table" piece.** More expensive to produce than the alternatives described above, a booklet usually includes lifestyle photographs as well as photographs of the home. Its advantages are its ability to project a high-quality image. Make sure the information in it has a long shelf-life.
- **Specialty and unique pieces.** These promotion pieces are the most expensive to produce because they use non-standard-sized paper and one-of-a-kind designs. They may have an unusual shape or feature a *die-cut* or *embossed* printing. The advantage, of course, is they help prospects remember your homes.

See Checklist 9.1, Brochure Guidelines, to ensure your brochures have all the elements for success.

Checklist 9.1 Brochure Guidelines

___ Use a typeface that is easily readable.

___ Don't use all caps or fancy script since they are harder to read.

___ Just as in ads, make generous use of white space to keep the design clean and uncluttered.

___ Colors, graphics, and design should match the style of your signage, sales office exhibits, and ads

___ Appeal to your target market. For example, for first-time buyers, a colorful, contemporary brochure would work well. For a luxury move-up market, a more elegant piece would work better.

___ Have a previous customer or two critique your brochure's appeal.

___ Make sure your brochure, flyer, or folder includes the following information:

    ____ Company name, logo, web address, phone and email contact

    ____ Builder's story or company mission statement

    ____ Photograph or rendering of your homes

    ____ Features and benefits

    ____ Location or map

    ____ Equal Housing Opportunity logo

Depending on your budget and the need for flexibility, your brochures can also include the following information, or you can convey these facts through inserts within a folder:

___ Price range of homes

___ Selection of floor plans

___ Options and upgrades available

___ Information on the community or communities in which you build

___ Lifestyle photographs of the target market

The following design options may cost more, but they will help distinguish your brochure from your competitors':

___ When you use black-and-white inserts, put them in a colored pocket folder.

___ Use a bold or distinctive color or add a varnish to make the color stand out.

___ Incorporate an unusual shape. Design a brochure in the shape of a house, for example.

___ Use a textured paper or card stock.

___ Use a non-standard size.

___ Be creative with die-cuts. Cut out a palladium window on the cover, for example.

___ Emboss, deboss, or foil-stamp some of the text and/or graphics.

___ Instead of or in addition to a paper brochure, a high-quality CD or DVD showing a virtual tour of your homes and community is a memorable take-home item.

Price, Benefits, and Features Lists

Some builders develop separate literature to list their homes' options and upgrades. You may want to omit prices from these lists to avoid frequent updating or revealing competitive information.

Site Map

If you are building on several home sites in one community, you want your prospects to be able to see on a map the sites that are still available and those that may have a lot premium attached. This can be a simple 8½ × 11 in. handout or an impressive, color-coded brochure, but keep in mind that the status of the home sites will change frequently. For that reason, an easily updated home site on a CAD program such as SmartDraw that can be reprinted often is a good option here. In larger communities, this is sometimes demonstrated via a topo table.

Community Information

This optional and separate flyer or brochure identifies area schools, provides average monthly utility rates, and indicates the locations of churches, stores, parks, airports, and cultural centers. It is particularly useful for buyers who are relocating and not familiar with the area. Your chamber of commerce may already have produced a brochure or relocation package for this purpose that you can purchase or use for free. Your area Realtors will likely have a relocation package that they use as well.

Supplier and Vendor Information

Suppliers and vendors sometimes produce materials about their products and services builders can pass on to their customers. They may include information on financing, appliance packages, smart systems such as security and lighting, and home warranty information. These collateral materials may have a place for you to stamp or sticker your company logo so you can customize them. Check with mortgage companies, title companies, utilities, cable television providers, building product suppliers, and school districts to see what printed information they have.

Write Effective Copy

What your brochure says is as important as how it looks. Follow the guidelines in Checklist 9.2, Copywriting Guidelines, to ensure your message is clear and effective.

Checklist 9.2 Copywriting Guidelines

___ Hire an ad agency or professional writer to write the copy.

___ Make the copy clear, crisp, and concise. Write it once, revise it, and carefully scrutinize the copy for repetitive or unnecessary words.

___ Explain the benefits of your homes, not just their features.

___ Summarize your points in bulleted lists for easy readability.

___ Use colorful and descriptive adjectives to paint a picture.

___ Keep the writing concise so readers can easily remember the main points.

___ Ask at least two people in your company to proofread the copy.

___ Be sure to run your copy through spell-check, although be aware that it won't catch every mistake (such as the difference between prospective and perspective or complement vs. compliment.)

Consider Costs

Although it's always best to get the same quality for less money, cost and quality are often intertwined. Nevertheless, you can curb production costs if you are on a tight budget. Here are some tips:

- Standard-size paper such as such as 8½ × 11, 8½ × 14, or 11 × 17 in. will cost less than other sizes, in most cases.
- Colored paper is usually more expensive than white in the same grade and category. Textured paper may be more expensive than smooth paper, depending on the brand and grade.
- Printing with black ink only is least expensive.
- Printing with black ink plus one additional color is the next step up in price.
- Printing with four-color process inks is the most expensive. This yields full color capability as in a color photograph.
- Special effects that call for a five-color press will add costs; for example, using a *varnish* on a four-color piece to protect the colors or to create a particular look.
- Pencil or ink renderings and line drawings are less expensive to reproduce than are screens or photographs.
- In designs that *bleed*, the graphic or photograph continues off the edge of the page. This special effect makes an ad more expensive to print than ones confined between the margins.
- Embossed, *debossed*, or *foil-stamped* designs cost more.
- Die-cuts add cost. You need a die-cut if a design is carved into the paper or if your brochure or paper is cut to an unusual size or shape.

Who Will Produce Your Collaterals?

Professionals who can help you produce your collateral materials include the advertising agency that produces your ads or a graphic designer. A graphic designer may have contacts at companies that can print your pieces as well. You may save 10 to 15% if you deal with the printer directly, but the savings may not be worth the time and trouble. If you are a small-volume builder on a tight budget, you may want to do all of the work yourself, but take special care to ensure your literature looks professional. Today's technology, including laser printers and new graphic design software, has made it easier for nonprofessionals to create professional-looking materials.

Producing Your Own Collaterals

If you have a good eye for design and some training in it, you might want to consider producing some or all of your own collaterals. Move forward if you have state-of-the-art graphic design software and are proficient with it. Some popular software programs that graphic designers and printers use are listed below. These programs are among the easiest to learn and will be compatible with equipment commonly used today.

- Microsoft Publisher. An entry-level application for page layout and design; targeted to the small business owner/entrepreneur market.
- Adobe InDesign. Suitable for creating posters, flyers, brochures, magazines, newspapers and books. Can also publish content suitable for tablet devices.
- QuarkXpress. Used by designers and large publishing houses to produce a variety of layouts, from single-page flyers to magazines, newspapers, and catalogs.
- Adobe Photoshop. A graphics editing program for manipulating photographs.

Consider having at least your primary brochure developed by a professional graphic artist. You can supplement the effort by making inserts, floor plans, and benefits and features lists yourself. Use inexpensive, matching papers for stationery, folders, flyers, and business cards on which you can print your own text and graphics. Use your own local source or contact PaperDirect at paperdirect.com.

Choose the Right Printer

- Get references from other professionals and ask for samples of the printer's work.
- Ask about their strengths and specialties. For example, a printer with only a two-color press will not be able to handle a four-color brochure without using a contractor. On the other hand, if you want a simple twocolor brochure, a printer with elaborate equipment might not be cost-effective.
- Find out if the printer will be doing the work from his or her own shop or contracting it to others. If they are contracting it, you may be paying extra or your job might not get enough oversight.
- Does the printer use the latest technology? Does he or she have the most commonly used software to read graphics? Refer to the previous list for some of the most popular programs.
- Distribute a standardized bid letter or request for proposal to all printers who are submitting bids to ensure fairness and consistent pricing, and get at least three bids to ensure you are getting a competitive price.
- Don't automatically award the job to the lowest bidder. You usually get what you pay for. Make sure a low bid did not result from a mistake.

Now that you have created effective collateral materials, you must also make an impact on the community at large. Turn to the next chapter to see how to cast a favorable light on your company name and keep it in the public eye.

PROMOTE POWERFUL PUBLIC RELATIONS

10

Public relations is the art of keeping your company name in the public eye and continually communicating the quality products and excellent reputation your company has. Every contact with your prospects, the real estate community, government officials, community leaders, trade contractors, and suppliers is an opportunity to enhance public relations. Word of mouth and reputation can make or break a builder. Therefore, professionally controlled social networking, positive media messages and constructive community involvement are keys to a builder's success.

Carefully Control Your Social Networks

Social networking has become a big part of any company's branding efforts. Social media provides an opportunity to enhance your online presence and to interact with your prospects and create connections with customers in a casual, conversational way. One or more people in your company should have specific responsibility to monitor and post to your social media outlets regularly. Below are the most important social networks, and their icons, on which to have a presence. Be sure to include the icons of the various social networks in which you participate in all of your collateral. Social media should not be relied upon alone; rather, it should be incorporated into an overall marketing strategy.

- **Facebook.** Allows users to post instant updates, photos, and news stories to a group of followers, called "friends."
- **Twitter.** Allows the user to post up-to-the-minute news items, called a "tweet" in 140 characters or less to his or her audience, called "followers."
- **Pinterest.** Allows you to "pin" pictures of designs, styles, housing elevations, floor plans, and housing products that you want others to view, comment on, and pass on to their Pinterest followers.
- **Houzz.** A repository of architecture, housing and design ideas. Use Houzz to showcase your work and to share ideas and communicate with your customers.
- **LinkedIn.** An professional directory that allows a participant to tout their talents and skills by way of posting a resume and job history. Professionally done, it can raise the profile of your company owner(s) and management to have individual listings on LinkedIn.
- **YouTube.** Your professionally-produced videos and virtual tours of homes, communities, amenities and surrounding area can be posted here with a link to your website. When users enter certain key words, your video will garner "hits."
- **Yelp.** A review-sharing site where customers of everyone from plumbers to shoe stores to home builders can post their experiences and give them a rating. If you have excellent reviews on Yelp, you can post the sign: They love us on Yelp!, available through Yelp.com.

For successful social media campaigns, refer to Checklist 10.1, Successful Social Media Guidelines.

Checklist 10.1 Successful Social Media Guidelines

___ Begin by securing your company name on Facebook and Twitter, even if you are not ready to launch your social media strategy.

___ Identify your target audience(s). What are they looking for in terms of content?

___ Define what you want to accomplish. Do you want to share community updates, PR stories, local market conditions, or a combination of all of those?

___ Link all social networking activities back to your company website.

___ Post frequently according to set goals, such as daily on Twitter and twice weekly on Facebook.

___ Monitor your social networking sites daily, several times each day is better, for any type of complaint or unfavorable post and address it immediately.

___ Avoid political or divisive posts. You will invariably alienate 50% of your audience.

___ Never use social media to disparage your competition.

___ Add the icons of the social media you use to your website, brochures, and email communication.

___ Be sure to post on your social media sites any time you are hosting a grand opening or other special event.

___ In larger firms, have restrictions in place as to who can post and what type of content they can post.

___ Remind your employees that their personal social networking sites are a reflection of your company as well.

Establish a Blog

One way to take control of your online marketing efforts is by maintaining a blog (short for *Web log*). One good online resource, called WordPress, provides customizable templates for do-it-yourself bloggers. Through a blog, you can provide regular and ongoing comments of interest to your audience. Regularly posting to your blog will provide content for search engines, which will drive greater traffic to your website.

> For more in-depth reading about social media, read *Social Media 3.0: Its Easier than You Think* by Carol L. Morgan, available from Builderbooks.com.

If you don't have the time or expertise to maintain a regular blog, consider hiring a professional content writer. Two resources are blogging.org and textbroker.com. Be sure to read each item yourself before it is posted. In Checklist 10.2, Successful Blogs, you will learn what you need to know to begin and maintain a blog.

Checklist 10.2 Successful Blogs

___ Define your audience. Are you blogging to prospects, customers who have already purchased, the public at large or a combination?

___ Determine what would be interesting to your audience in terms of content.

___ Post consistently, at least every week.

___ Make sure your blog links back to your website and to your social media sites.

___ Offer insights or information that will be relevant and of real benefit to your audience. Readers will return to your blog if they find a value in it.

___ Avoid political or divisive posts.

___ If your blog site enables readers' comments, monitor your blog every day for unfavorable comments so that you can address them immediately.

___ If you hire a professional writer to provide content for your blog, be sure to proof it before it is posted.

Tip from a Pro

Meredith Oliver, an expert on Internet marketing strategies and president of Creating WOW Communications, recommends asking the following questions to see how well you are integrating social media into your overall marketing plan.

- How often are you posting on your business Facebook page?
- How often do fans like, comment or share one of your posts?
- Is the fan base growing steadily on your social media sites?
- What sites beyond Facebook are you using? YouTube, Twitter, Pinterest?
- Do you have a blog that is tied into your social media sites and is optimized for Google search?
- Are you using email marketing to drive fans to your blog and social media pages?

Used with permission from Meredith Oliver, MIRM, President, Creating WOW Communications.

Tech Tip

Sign up for Google alerts on your company name, community name, your owner(s), executives and branded products and services.
You will be notified when any of these key words in published, for example, in a negative review on Yelp.

Send Press Releases to the Media

One of the best ways to get free coverage in the media is to send regular press releases for all newsworthy events. Make a habit of sending them, and if they are done well, inevitably you will get coverage. Send press releases whenever you do the following:

- Open a new model for viewing.
- Build on homesites in a new community.
- Add new floor plans to your inventory.
- Release a new phase of homesites for sale in a community.
- Hire new personnel.
- Promote an employee to a new position, particularly if that employee will have contact with the public; for example, "ABC Builders Announces Joanne Williams as New Customer Service Coordinator."

- Present an employee with an award, certificate, or other honor.
- Announce a grand opening or special promotion.
- Announce new incentives or special pricing.
- Develop an innovative design or meet a unique need of a market; for example, "Well-Built Homes Introduces Custom Homes with Hangars on the Air Strip."
- Profile a satisfied customer.
- Gather interesting information through your market research, such as "Study Shows Silicon Valley Buyers Demand Wireless Home Offices."
- Use unique or state-of-the-art design, construction or energy-saving techniques, such as "ABC Homes Brings Zero-Energy Homes to Austin."

Write Effective Press Releases

Many press releases are sent to the media via email. Some prefer to send a hard copy through the mail. For hard copy releases, or for releases sent as pdf attachments, follow the guidelines in Checklist 10.3, Effective Press Releases. A sample press release is shown in Figure 10.1.

Checklist 10.3 Effective Press Releases

___ Use double spacing. Editors will want room for their own notations.

___ Include the words "FOR IMMEDIATE RELEASE" at the top if the editor receiving the information can publish it right away. If not, include a release date on which the media may publish the information with the words: "Embargoed until December 19th, 20xx."

___ Include in the heading a contact name with phone number and email address, so the editor can obtain additional information.

___ Begin the release with a descriptive and attention-getting headline.

___ Organize your release with a *lead,* or introductory paragraph; a body; and a concluding paragraph, in that order. The lead paragraph should answers six questions: "Who?" "What?" "When?" "Where?" "Why?" and "How?"

___ Write the remaining information in descending order of significance. (If an editor chooses to shorten the release for space needs, it will often be cut from the bottom, so include the most important information at the beginning.)

___ Always include a quote or two from the company principal or spokesperson.

___ Include photographs if possible.

___ If the press release is more than one page, indicate that by typing "more" at the bottom of each page except the last one.

___ Type three hashtag symbols at the bottom of the last page, to indicate the end of the release, e.g., # # #

Figure 10.1 Sample Press Release

[Builder's Letterhead]

FOR IMMEDIATE RELEASE
April 15, 20xx
Contact: Susan Jones, ABC Builders
Phone: (801)-555-4567
sjones@abcbuilders.com

ABC Builders Features Cookie Monster at Grand Opening Celebration

Phoenix, Ariz.—In a grand opening celebration for people of all ages, ABC Builders is hosting a gala event from noon to 4 p.m., on Saturday, April 27, to unveil its two newly furnished models at Desert Breeze in Phoenix. On hand to greet guests will be the Cookie Monster himself, and he will be handing out chocolate chip cookies and coloring books to ABC's littlest visitors. All visitors will be eligible to register for a drawing for a trip for four to New York City to see a taping of Sesame Street.

The grand opening also will feature balloons, face-painting, and a playground to occupy the children while their parents peruse the models. ABC Builders offers Pueblo style, three- to four-bedroom floor plans for growing families. The homes feature stucco and stone exteriors, beamed ceilings, oversized closets and pantries, kitchen islands, covered porches and decks, and high-speed Internet connections.

The firm has been offering homes in the Phoenix area since 1982 and enjoys a first-rate reputation as a high-quality home builder with affordable prices and superior customer service. Prices at Desert Breeze start at $150,000. Green-belted jogging and hiking trails lead to two neighborhood parks.

ABC Builders is the only local builder offering homes in the Desert Breeze community. To get to Desert Breeze, take I-10 to the Chandler exit, and go west 2½ miles to Saguaro Street. Turn left and then follow the signs to Desert Breeze.

###

Sponsor Community and Charitable Events

Scout out your city or area for opportunities to sponsor community and charitable events. You can benefit the community and also familiarize people with your company name. Below are some suggestions of events you may want to sponsor:

- Sponsor a local Little League or youth soccer team and provide the team with T-shirts with your company name on them. This activity is ideal if you build homes for young families.
- Become involved in other youth-serving groups like Boy Scouts, Girl Scouts, Big Brothers, or Big Sisters.
- Volunteer with charitable groups focused on housing, such as Habitat for Humanity.
- Cosponsor a 10K run, a walk-a-thon, bowl-a-thon, kids' bike-a-thon, or other event that raises money for charity.
- Join your local chamber of commerce and/or economic development council, and have a representative participate in its activities.
- Join your local board of Realtors® and have a representative attend its functions.
- If you have a conference room in your administrative office or sales office, offer the meeting space to a community organization.
- Sponsor a portion of a highway in a litter cleanup program. In exchange, your company name will be posted on a sign and exposed to hundreds of drivers. (Make sure you are prepared to maintain that stretch of road!) Call your state's department of transportation for more information.

Network with Community Leaders

You should regularly network with the leaders in your community. In a large company, the officers and sales representatives should become involved as well. Building rapport and cultivating goodwill with the following groups makes doing business more enjoyable and can yield referrals:

- School officials in the districts where you build
- City council members
- Community business leaders
- Board members of the homeowner associations in communities where you build
- Trade contractors, suppliers, and vendors
- Members of the media
- Key real estate agents in your community
- All real estate agents with whom you have cooperated or who have registered at your sales office, model, or inventory home

Here are some suggestions for networking opportunities:

- Volunteer to speak, participate in seminars, or give presentations, and serve on local or national committees for your local HBA, SMC, board of Realtors®, regional, national, or international builder conventions, or local organizations such as the Lions Club or Executive Women's Club.
- Network with and maintain positive relationships with local reporters and editors. Be helpful in providing quotes but make sure you choose your words carefully.
- Make company officials available for interviews on topics such as how to choose a custom builder, energy efficiency, earthquake preparedness, financing, etc.
- Contact local radio talk shows that may deal with business, finance, or real estate issues, and offer to appear as an expert guest.

Learn to be a Spokesperson

When representing your company to the media, be sure to respond to all inquiries immediately. Remember that reporters are on a deadline and if you miss it, you'll miss the opportunity to tell the story your way. The Public Affairs department of the NAHB offers spokesperson training. Two courses from NAHB teach you valuable skills when representing your company to the public and the media.

- Spokesperson Training: Question and Answer Skills for Business & Media—Teaches how to conduct media interviews and respond to questions with confidence and skill.
- Spokesperson Training: Presentation Skills—Reviews message development for the target audience; body language; engaging your audience; and organizing a presentation.

For more details on these courses, as well as where and when they are offered, see NAHB.org/education.

Develop a Newsletter

A company newsletter provides an excellent way to deliver a regular and consistent message. Send by email or snail mail at least three times a year, but you can make it quarterly, bi-monthly, or even monthly. Your mailing list should include the following people:

- Previous customers (They will remember you when they need to move up or move down, or they may refer you to a friend.)
- All prospects (gathered from your customer relationship management program).
- The real estate community (You can hand-deliver copies to key real estate offices near you to obtain another opportunity for contact.)
- Residents in rental communities located near your building sites
- Local chamber of commerce members
- Local newspaper editors
- News editors at local radio stations

Make Your Newsletter Stand Out

The tips in Checklist 10.4, Creating a Newsletter, will help you ensure that recipients will read and remember your newsletter.

Checklist 10.4 Creating a Newsletter

___ Design your newsletter to remain consistent with your overall branding, with similar color, design, and style as the graphic elements in your collateral materials.

___ Create a name that is memorable, easy to pronounce, and in keeping with your local culture, for example, *The Sand Dollar* for a Jacksonville, Florida builder or *The Grapevine* for Napa, California builder.

___ Incorporate the name in a *masthead* that is attractive, eye-catching, and ties in well with homes or home building.

___ Include company news such as new model openings, new personnel, grand openings or special events.

___ Include news and feature items from third-party sources so your newsletter is informative as well as a good public relations vehicle for your company. Search local and national industry publications for interesting and relevant stories to include. Obtain permission in writing from the publisher for any material you want to reprint.

___ Include photographs, floor plans, elevations, other drawings, cartoons, or other graphics as often as possible.

___ Include maintenance tips as appropriate for the season.

___ For paper newsletters, print with at least two colors (commonly black plus one other color) and make sure the color contrasts well with the paper for readability.

___ Use a simple font. *Serif* fonts are preferred because they help guide the eye from one letter to the next. On the other hand, elaborate scripts are hard on the eyes.

___ To keep production costs as low as possible, page counts should be divisible by four. (An 11 × 17 in. sheet of paper folded in half makes a four-page newsletter. Fold in half again or in thirds to mail.)

___ Produce your newsletter as a self-mailer to reduce labor and printing costs. Keep one-half to one-third of a panel of the newsletter blank for your return address and the address label (Check with the U.S. Postal Service for specific requirements before you print.)

___ Include the company name, telephone, email and website address.

___ Include social media icons on which you are represented.

Consider Using Advertorials

Advertorials are articles provided by an advertiser to a publication and packaged to look like news. You'll find these often in a real estate section or supplement or new homes guide. You can write advertorials to get exposure in the media. They are similar to press releases, but they differ in important ways. Some publications charge money to guarantee advertorial placement. Others are eager to fill up editorial space, so they don't charge. Some publications may accept advertorials as a trade-off for a certain amount of purchased advertising. Ask the editorial staff of your local newspaper and magazines what their policy is for advertorials. The following will help you distinguish a press release from an advertorial.

Press Releases vs. Advertorials

Press Releases

- Placed as space allows or is at the discretion of the editor
- Are written from the viewpoint of the submitting firm or its agent.
- Placement is free. They incur no fees.
- Will not include the heading: Advertisement.
- Often are shortened, edited, or rewritten by the publication's editor.
- Standard industry practice and accepted by almost every publication

Advertorials

- Usually have guaranteed placement
- Appear to read as an article
- May incur a fee or may be contingent upon a certain amount of paid advertising.
- Publication may require the inclusion of the heading, "Advertisement."
- Submitting firm has more editorial control than for a press release.
- Some publications do not accept them, but a few publications carry primarily advertorials. Advertorials in these publications may attract proportionally fewer readers than those in newspapers.

Customer Testimonials

Some of the most credible voices for extolling the benefits of your homes come from satisfied customers. Below are some of the ways you can share your positive testimonials:

- Post them on your Facebook or Twitter page (with your customers' permission).
- Feature a Home Buyer of the Month column in your social media posts or newsletter. Profile individuals, couples, or families, and include their comments and photographs.
- Keep a binder in your sales office, model, or speculative home with thank-you notes or favorable letters that you have received from customers.
- Use positive comments from customers' responses in your ads, brochures, newsletters and on your website. (Be sure to get their written permission first.)
- Mount and frame several of the best testimonial letters you have received and display them in your sales office, model, or speculative home.
- Learn who your customers are. Many of them are interesting people who could be featured in your newsletters; for example, "Mom-and-Daughter Musical Duet Moves in at Centennial Heights," or "ABC Builder's' Customer Named School Teacher of the Year." If you approach your customers about profiling them, they are likely to provide positive quotes about your homes as well.

You can enhance your public relations program with special promotions. The next chapter will provide ideas for planning effective promotions, including direct mail campaigns.

PLAN EFFECTIVE PROMOTIONS AND TARGETED DIRECT MAIL

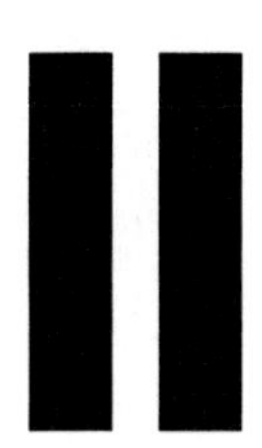

To generate excitement, increase traffic, and create a sense of urgency, conduct occasional special promotions, *email blasts,* and direct-mail campaigns. Promotions are limited campaigns tied to a one-time, on-site event or a special program, prices, financing, or other incentive. Email blasts, text marketing and direct mail campaigns can be used to supplement and publicize these events or they can stand alone to bring traffic to your site. The following guidelines will help you plan, advertise, and execute effective promotions and direct-mail campaigns.

Grand Openings

One of the most common and well-known on-site builder promotions is the grand opening. A grand opening can celebrate a community opening, a new phase, or models or inventory homes opening for touring. Consider these variations on the grand opening to provide additional opportunities to draw traffic:

- **Pre-grand opening.** Conduct this event one or two months before the actual grand opening and before you have a completed model or inventory home to show. Use it to generate sales prior to construction.
- **Re-grand opening.** Sometime after the initial grand opening this event could reestablish awareness and generate additional excitement. It is particularly helpful if traffic has slowed or you have refreshed your models.
- **Customized grand openings.** Create two or more grand openings for different audiences. For example, one for the real estate community and the press, and another for the general public. You could tout this as a "VIP Grand Opening."

Other On-site Promotional Ideas

Listed below are a variety of other ways to bring traffic to your site. You can tailor your theme and modify your activities according to your particular target market, your specific objectives, local customs, climate, and of course, your budget.

Conduct Seminars

Host an educational seminar or series of seminars in your sales office, model, or inventory home. Contact local professionals who may be willing to share the cost and the opportunity for publicity. Listed below are some types to consider:

- Conduct a home buying seminar with speakers representing a mortgage company to discuss financing, a title company to cover closing consideration, the local chamber of commerce to talk about area amenities, and of course, your building company to discuss construction and energy-efficient techniques that make your homes superior.

- Ask your merchandiser or local designer to feature a decorating seminar, with do-it-yourself tips on window treatments or wall treatments.
- Your landscape architect could conduct a workshop on landscaping and gardening tips appropriate to your soil and climate. Focus on native plants for any area and *xeriscaping* in desert or drought-prone climates.

Celebrate Holidays

Plan for only one or for a calendar full of seasonal activities. Special occasions provide you with a built-in theme when people are already in a festive mood.

- **Host a holiday house.** In an appropriate market, cooperate with one or several local designers to decorate one of your models or inventory homes with holiday trimmings. Also keep the house open during evening hours when lights attract attention and the competition is closed.
- **Plan a special Trick-or-Treat event.** On-site sales representatives can pass out treats to young children throughout the model home, and you can involve them in traditional Halloween/harvest activities such pumpkin painting and bobbing for apples. To reduce the mess, try hanging the apples on strings at different heights, rather than using a tub of water.
- **Have an egg hunt.** Hide eggs on the ground surrounding your sales office, model, or inventory home, and feature a live bunny to greet small guests. Selected eggs could contain prizes the whole family can enjoy, such as gift certificates to local restaurants or retailers.
- **Celebrate Memorial Day, Labor Day, and the Fourth of July.** Plan traditional picnics, with hot dogs, apple pie, and soft drinks. Display the American flag, and dress your sales staff in red white, and blue. If your budget, local laws, and insurance coverage allow it, present a fireworks display at dusk. If you are participating in a multibuilder community, invite other builders to cosponsor the fireworks and to contribute to the cost. (Check with your local fire department for local ordinances and arrange for on-site supervision. Also check with your insurance agent for single-event liability coverage if it is available.)
- **Observe Mother's Day.** Offer a free luncheon buffet for guests who bring a mother with their party to tour your home. Pin a corsage on the mother. Register all moms for a drawing for a day of pampering at a local salon.
- **Remember Father's Day.** Serve barbecue and pass out caps or T-shirts with your logo to the fathers. Register dads for a drawing of new golf clubs, fishing equipment, or an iPad.
- **Acknowledge odd holidays and events.** Celebrating odd "holidays" can draw attention because they are unusual and fun. For example, Ground Hog Day or National Pickle Week. Celebrate creatively to reflect the theme.

Plan Your Promotion

When it's time to launch your on-site promotion, you'll want to make sure that the event is smooth and successful. Preplanning will pay off. You may want to enlist the services of an event planner who can coordinate all aspects of your event, including invitations, entertainment, and food. For guidance on who to invite, how to coordinate the details, and how to staff your promotion or special event, refer to Checklist 11.1.

Checklist 11.1 Planning Promotions and Special Events

Your Invitation List

- Send hand-addressed invitations to the following people:
- Chamber of Commerce members

- School officials in the districts in which your community is opening
- City and local government leaders
- Community business leaders
- Officers of the homeowner's association in the community in which you are opening
- Residents of nearby rental communities (if your target area is first-time buyers or young professionals)
- All prospects from your visitor registrations cards/CRM
- Current homeowners
- Key real estate agents who have registered at your site or with whom you have cooperated on a sale in the past
- Members of the press

Work Out the Details

The following checklist will aid you in your on-site promotion:

- If your event includes touring a model or speculative home, make sure it is complete, immaculate, and ready for touring. Make sure the carpet is installed and that all furnishings, vignettes, and accessories are properly placed.
- Set up a reception table in order to greet guests and register every visitor. Use the registration as a chance for the guest to win a prize, and keep the names to use as a database for future mailings and prospect follow-up.
- Tie the event to a theme, such as a beach party, clam bake, luau, or a "Spring Fling."
- Plan for food and drink and mention it in your advertising for the event.
- Conduct a drawing for a grand prize. Tie the prize with the theme, for example, a beach party could involve a trip for two to Ft. Lauderdale, or for smaller budgets offer a series of less expensive gifts: a new pair of designer sunglasses or free admission to a local water park.
- Have plenty of brochures and collateral materials printed beforehand. Make sure visitors take information with them.
- If your budget allows, give every visitor a give-away with your name and logo on it. You could give coloring books to the children, and caps, T-shirts, mugs, or even specialty candies to the adults.
- Advertise the promotion in newspaper and radio ads, in your newsletter and on your website. Start 3 weeks ahead of the event. Increase frequency closer to the date. Consider a radio remote to broadcast live updates to the public.
- Erect directional signage at nearby major intersections.
- Feature on-site attractions that are easily spotted by drivers-by, such as a colorful tent, balloons, and even spotlights if it is an evening event. (Evenings will draw single professionals, dual-income couples, and families.)
- Plan for traffic control and parking. If you expect a large crowd, provide ample parking, perhaps even attendants to valet park or direct visitors to the closest available spots.
- Choose a time of the year when the weather is most likely to cooperate. Have a contingency plan in case of rain, including a back-up date and a tent under which to take refuge.
- Consider cooperating with one or more other builders in a community in which you build to share the costs of an on-site promotion.

Staff Your Event

The following specifications will help you plan for adequate personnel coverage during your special event:

- Two people to greet and register guests
- One to three people to park or direct cares
- Two to four sales professionals, real estate agents, or sales assistants

- One emcee to formally greet guests, make announcements, and conduct drawings. (You could fill this role or your marketing director, sales manager, or a local radio personality could do it if you hold a radio remote.)
- A caterer or three to five people to serve food. Other specialists as appropriate to your particular event, such as a clown or face painter for a market with young children, a chef for gourmet cooking demonstrations for a move-up market, or a band to play music.

Hold Special Events for Real Estate Agents

You may want to conduct a separate grand opening or other special on-site event strictly for the real estate community, especially if any of the following statements are true:

- You are trying to foster greater participation with the real estate community.
- Your area has a large number of real estate agents.
- Broker cooperation comprises a large portion of your sales.
- You rely solely on real estate agents to sell your homes.

The following are suggestions for hosting a separate event for real estate agents:

- Hand-deliver invitations to all area real estate offices and place them in the agents' in-office mailboxes.
- Speak to those agents in person who happen to be in the office at the time.
- Personally hand-address invitations to specially targeted agents including those who are top producers in your community, those who have brought and registered customers at any of your model or inventory homes, and those who have cooperated with you on a past sale.
- Ensure that brochures or flyers available during the event include information on broker registration and cooperation and that they clearly outline your policy for broker cooperation.
- Offer food and drink. A luncheon or dinner event works well for busy real estate agents.
- Offer a prize relevant to the real estate profession, such as a free cell phone or a coupon for car detailing.

Tapping into a Hot New Trend

With gourmet food trucks being all the rage in L.A., the developers of TLofts, a new loft project in West Los Angles, invited a variety of gourmet food trucks to visit their location every day at lunchtime. Feeding almost 7,000 nearby office workers gained attention on Twitter and Facebook as a regular food truck stop with TLofts as a backdrop. As a result, TLofts has become an LA destination and a darling of the blogosphere.

Used with permission from Carol Ruiz, NewGround PR& Marketing.

Budget-Cutting Tip

Ask a local high school or college to provide a jazz band for your grand opening or on-site promotion.

Special Incentives

Some promotions may involve special prices, financing, or other incentives. Advertise on your website, through social media, in the newspaper and on the radio and plan to have additional sales representatives or assistants while the promotion is active. Below are some inspirations to help you come up with special incentives that will work for your targeted buyer:

- "Buy this weekend and we'll throw in an in-ground pool for free."
- "Get 2.5% interest rate this Sunday only!"
- "Buy this weekend and receive one year's free membership at Oakland Athletic Club."

- "Reserve your home site during our grand opening and get $10,000 in free options and upgrades."
- "Buy a home from ABS Builder this week only and get the high-tech home office option for free!

Parade of Homes

You can gain wide exposure in your community by participating in a Parade of Homes. Similar events are sometimes called "Street of Dreams,""Kaleidoscope of Homes," or "Spring Tour of Homes." The local HBA usually plans and administers these events. Participation in events like these requires a large investment of time, but will yield media attention and buyer referrals. See if your local HBA hosts such and event and ask how to participate. The event requires a great deal of planning and coordinating with other Parade participants and an oversight committee.

- Generally, here's how it works:
- Several builders (as few as 6 to as many as 500) provide a fully furnished home for the public to tour.
- The homes can be built on scattered sites throughout the area or together in one community. Participating builders usually pay an application fee that may be nonrefundable.
- Attendees usually pay admission fees, some or all of which may be donated to a charity.
- Local merchants often donate furnishings and accessories in exchange for publicity.
- The parade event usually features a contest with awards such as Best in Parade and Best Interior Design. Winners benefit from the related publicity
- Participating builders should resist the urge to over-impress the public just for the parade event. Use features, furnishings, and accessories that will appeal to your target market.

For tips on how to best prepare your Parade home for public viewing see Checklist 5.3, Excellent Exteriors and Checklist 7.2, Model Furnishings and Accessories. For more information on Parades of Homes contact your local home builders' association or sales and marketing council.

Email Blasts and Direct Mail Campaigns

Email blasts, regularly sent email messages to your prospect list, also known as a "drip campaign," and direct mail can attract traffic to your homes. You can use these types of campaigns to promote a special incentive, offer, or to keep your name in the public eye.

Where to Get Mailing Lists

Contact a market research firm or mailing list service or broker. These companies often have not only area-specific lists, but also lists that categorize households according to demographic and *psychographic* information, occupation, and lifestyle. The following nationwide companies can help you with lists and advice on direct mail:

- Leadsplease.com
- Lists.Nextmark.com
- Mailinglist.org

Who to Include

To develop a direct mail or email database, you can purchase address lists from a mailing house. Or you can compile names and addresses from the tax records in your local county clerk's office. If you have a relationship with a title company representative, he or she may be able to obtain this list for you. These documents may include information such as how long residents have owned their homes, home size, and purchase price. If you want to narrow your list based on specific types of buyers, below are some ideas of how to do this:

- To attract first-time home buyers, send direct mail pieces to rental communities within a certain radius of your homes.
- If you are selling water or recreation properties, get a list of registered boat owners from the appropriate state licensing agency.
- If you are building high-end custom homes or homes in a golf course community, see if you can buy a mailing list from a country club or golf course.
- If you build for seniors or retirees, ask for a list of grandparents.
- If you specialize in first-time move-up buyers, request a list of new parents.
- If you offer homes in a country club community, ask for golf or tennis fans.
- If you build in a resort or "snowbird" community, ask for people who have inquired about ocean or river cruises.

Three Methods of Compiling Mailing Lists

There are three main ways you can compile street and email addresses for your direct mail campaigns.

- **Purchase a list.** You may select a group based on demographic or geographic criteria. For example, "Pleasanton residents with children". You own the list and can use the names as often as you want.
- **Rent a list.** This option is less expensive, but you don't actually see the addresses and must work with your provider to send the content.
- **Generate your own list.** These are people who have voluntarily given you their contact information, either through registering on your website or when they visited your model home or sales office. These can also be cultivated from your customer relationship management program.

Note: Be sure to always include an opt-out feature of any email campaign.

Ensure a Successful eBlast or Direct Mail Campaign

For successful eBlast or direct mail campaigns, see Checklist 11.2.

Checklist 11.2 Successful eBlast or Direct Mail Campaigns

Make sure your direct mail campaign includes the following items:

___ **Call to action.** Create a sense of urgency that requires prospects to act, as in the following examples:

- "Bring in this coupon to receive $1,000 worth of upgrades for free."
- "Show this coupon on your smart phone when you visit Livemore Lofts to be eligible for a trip for two to Hawaii."

___ **Appeal to the target market.** Although a first-time buyer might respond to a fun headline on a colorful postcard, you might need to include a gourmet chocolate or a champagne split for a move-up prospect for an upscale or resort community.

___ **Separate campaign for agents.** For the real estate community, you may want to conduct a separate campaign with its own special incentive, such as, "$500 bonus commission on first ten Realtor-referred homes sold."

___ **Repetition.** For maximum *return on investment (ROI),* include a budget for multiple mailings. Campaigns are most effective when direct mail recipients receive three mailings.

APPEAL TO THE GROWING MULTICULTURAL MARKETS

12

Increasing diversity and immigration in America has opened new opportunities for builders. The 2011 Census data show that the percentage of people who speak a language other than English at home is 20.3%.[14] Of the American labor force, a great many are immigrants. New immigration plus births to immigrants added more than 22 million people to the U.S. population in the last decade, and in March of 2011, the share of working-age (18 to 65) immigrants holding a job was the same as natives—68%.[15]

And they're coming from all over the globe, with the largest percentage increase in the last decade was from these countries: Honduras (85%), India (74%), Guatemala (73%), Peru (54%), El Salvador (49%), Ecuador (48%), and China (43%).[16]

The five states where immigrants make up the largest percentage of the population are, in order, California, New York, Texas, Florida, and New Jersey.[17] If you are building homes in any of these states or in other areas where buyers are coming from outside the United States, you would do well to educate yourself on the buying habits and preferences of diverse customers.

Be sure to research the immigration patterns and statistics in your specific area. Consult the U.S. Census Bureau at census.org, and search for State and County Quick Facts.

Marketing and Advertising to Multicultural Consumers

Targeting your marketing efforts to people from specific cultures requires the utmost sensitivity. Says S. Robert August, MIRM, of North Star Synergies in Denver, Colo., "The melting pot is so great today; builders should definitely be expecting a multicultural dimension in their business. The most important thing is to be respectful of everyone, never judging anyone by the way they look."

Another important caveat is never to assume that a customer from a particular ethnicity wants to live in an area with others of the same ethnic background. Be careful not to stereotype anyone based on his or her name, appearance, speech, gender, or any other characteristic.

Adhere to the Federal Fair Housing Act (see Chapter 8, Develop Your Advertising Program). At the same time, you can use advertising to demonstrate that you welcome any and all prospects. When you are creating ads to appeal to a specific demographic in your market, have experts review them.

The Asian-American Market

Asians were the nation's fastest-growing race or ethnic group in 2012.[18] This situation presents new housing demands and great market opportunities for builders. Understand that the "Asian market" represents a large and diverse group in itself, with origins from China, South Korea, Vietnam, the Philippines, and Japan.

Feng Shui

Builders who hope to capture the Asian market should learn about *feng shui* (pronounced "fung shway"). Feng shui literally means "wind-water" in Chinese. It refers to the orientation or configuration of sites, structures, or objects to harmonize with the forces that inhabit them. Some believe these energies can enhance one's health and fortune. Many people of Asian descent, particularly Chinese, believe, at least in part, in the fundamentals of feng shui.

For years, many home builders, particularly in markets where there is a large percentage of Asian buyers, such as Orange County, Calif., have hired feng shui consultants before they design floor plans, orient homes on a site, assign numbers to addresses, or merchandise model homes.[19] Many home builders have discovered that even for non-Asian buyers, adhering to feng shui principles result in better design.

Read more about it

To learn more about feng shui design principles, including locating consultants and courses, consider the following sources:

- *Feng Shui Do's and Taboos* by Angi Ma Wong
- *Feng Shui For Dummies* by David Daniel Kennedy and Grandmaster Lin Yun
- *Feng Shui that Makes Sense—Easy Ways to Create a Home that Feels as Good as it Looks* by Cathleen McCandless
- The American Feng Shui Institute (amfengsui.com)

Although there are many nuances to feng shui design principles, the tips in Checklist 12.1 will help ensure that you cover the basics.

Checklist 12.1 Feng Shui Do's and Don'ts

Homesite and Landscaping

Do

- Site a home so the back faces nearby mountains or an ascending hill.
- Choose home sites with shade trees in the backyard.
- Plant an abundance of green and glossy plants.
- Install fountains or pools to create the sound of running water.
- Be aware of the power of numbers in Asian culture when requesting addresses from your municipality; in general, avoid the number 4 and favor the number 8.

Don't

- Site a home so the front of the house directly faces mountains or an ascending hill.
- Site a home adjacent to or in front of a church, hospital, funeral home, or cemetery.
- Plant trees in a direct line with entrances, particularly the main entrance.
- Design sidewalks and walkways in straight lines; create meandering designs instead.

Stairways

Do

- Create curved staircases rather than straight ones.
- Locate stairs on the left side of the home as one enters.
- Design wide staircases and light them well.

Don't

- Locate a stairway in line with a front door.
- Locate a stairway at the center of the home

Bedroom

Do

- Place beds where they have a full view of the bedroom door.
- Place beds diagonally opposite the bedroom door.
- Position beds against a solid wall, not a window.

Don't

- Position a ceiling fan over a bed
- Position a bedroom so it is aligned with the doorway of another bedroom in the home.
- Use exposed beams in the master bedroom or over any bed.
- Locate the foot of a bed in direct alignment with the front of a bedroom door.
- Position a bed so that a bathroom is located directly to its side.
- Install mirrors so they reflect the occupants of the bed.

Bathrooms

Do

- Paint the room a soft, warm color.
- Keep drains covered and toilet lids down.
- Add candles (use flameless in your models) and soft-colored linens when merchandising or vignetting,

Don't

- Locate a bathroom so the toilet is visible from the bedroom door.
- Separate two bedrooms with a bathroom.
- Locate a bath or powder room beyond the front entrance of the house.
- Locate a bathroom above the main entrance of the house.

Sales Offices

Do

- Align the sales associate's seating at right angles with customers' seats.
- Include plenty of live plants and natural lighting.

Don't

- Place customer seating where the customer's backs will face the office door.
- Align seating where customers will sit directly opposite sales associate.

Other Design and Merchandising Tips

Do

- Design square, round, or rectangular homes.
- Front doors should be blue (for water) or black (representing career).
- Design open courtyards in the center of the home.
- Use red, blue, and green in decorating, except in the case of a Korean market, where you should avoid red.
- Use generous amounts of live, green plants.
- Use warm, rich colors on walls.
- Use spring, summer, or winter colors in decorating.
- Use mixed-ethnic group/team lifestyle shots of children, e.g., Scouting, soccer, baseball.

Don't

- Use dark colors on walls.
- Use all white in decorating.
- Use exposed beams anywhere in the home, especially in bedrooms.
- Position a fireplace in direct alignment with the main entry.
- Place a bathroom, staircase, kitchen, stove, fireplace, or skylight at the center of the home.
- Divide the house in half by a hallway.
- Create sharp corners in walls or countertops.
- Design L- or U-Shaped homes.
- Create long, narrow hallways.

The Hispanic Market

Hispanic or Latino? These two terms are often interchangeable and often confusing. For purposes of gathering U.S. Census data, Hispanics or Latinos are those people who classified themselves in one of the specific Spanish, Hispanic, or Latino categories listed on the Census 2010 questionnaire—"Mexican," "Puerto Rican", or "Cuban," as well as those who indicate that they are "another Hispanic, Latino, or Spanish origin." People who do not identify with one of the specific origins listed on the questionnaire but indicate that they are "another Hispanic, Latino, or Spanish origin" are those whose origins are from Spain, the Spanish-speaking countries of Central or South America, or the Dominican Republic. The terms "Hispanic," "Latino," and "Spanish" are used interchangeably.[20] In this book, we will use the term Hispanic to remain consistent.

The Growing Hispanic Population

In 2011, the percentage of the American population who identified themselves as Hispanic or Latino was 16.7%, and is expected to more than double, from 53.3 million in 2012 to 128.8 million in 2060. Consequently, by the end of the period, nearly one in three U.S. residents would be Hispanic, up from about one in six today.[21]

Characteristics

Builders catering to Hispanic markets should be aware of the following general characteristics that may affect demand for housing:

- There tend to be more people per household in Hispanic versus non-Hispanic households. Multiple generations commonly live together.
- Hispanics are more likely than non-Hispanics to be concentrated in geographic areas.
- This market is likely to provide customer referrals if satisfied.
- Religious faith often plays a strong role. Although more than 90% of the Spanish-speaking world is Roman Catholic, the Hispanic share of other churches, such as evangelical Christian, is growing.
- Hispanic people living in Texas or California tend to have conservative or traditional tastes, while those in Florida tend to have more urban or contemporary tastes.

Strategies and Designs

When designing and merchandising for the Hispanic market, consult architects and merchandisers familiar with those markets. Also consider hiring sales associates fluent in both Spanish and English. Consult your local HBA or SMC and network with local Hispanic leaders to help you find experts with this demographic. You may also contact the National Society for Hispanic Professionals at nshp.org.

Listed below are suggestions for marketing to the Hispanic market in general but you should become familiar with the desires of buyers in your local areas or the places where you sell homes.

- Design large eat-in kitchens.
- Design formal areas for entertaining.
- Build plans with four to five bedrooms.
- Include a downstairs bedroom-and-bathroom suite for visiting relatives or live-in parents.
- Advertise in both Spanish- and English-speaking media.
- Provide product and financing information in Spanish as well as English.
- Merchandise outdoor areas, such as patios, decks, and courtyards with outdoor furniture and built-in barbeque pits.

- If you choose to appeal to spirituality in any model home, do it subtly with candles and artwork, rather than overtly with religious symbols or other direct references to religion.
- Use warm, bright colors in model homes.
- Place framed photographs of multicultural "friends" and "family" throughout a model home.

The African-American Market

The U.S. African-American community represents a growing marketing opportunity for builders. The buying power of this demographic has surged over the past years, significantly surpassing the growth rate of the white U.S. population. In fact, the buying power of African Americans is predicted to reach over $1 trillion by 2015.[22]

One in ten African Americans live in the West and the South, and there has been a significant shift to the suburbs. By 2015, more than half of African-American households will live in a metropolitan suburb.[23] Also increasing are mixed-race families, Census data show. Married couples of different races and ethnicities grew by 28% in the decade between 2000 and 2010, from 7% to 10%.[24]

Things to keep in mind when targeting the African American market are: to be aware that this demographic group has been portrayed inaccurately and negatively in the past in media and entertainment; to remember that this group is bi-cultural, straddling the culture of mainstream America, as well as the African American culture; and like most ethnic and cultural groups, this group is actually composed of several submarkets and subcultures.[26] A good rule of thumb for marketing to any culture is to not paint any group with a broad brush.

> In a move that showed increasingly greater percentage of and tolerance for mixed-race couples and families, in the summer of 2013, General Mills ran a first-for-its-brand Cheerios commercial featuring a white mother, an African-American father, and a bi-racial child. Advertising executives for the company cited an intention to create a more authentic and true picture of the real American family today.[25]

PROVIDE SUPERIOR CUSTOMER SERVICE

13

When does customer service begin and end? Service-oriented builders know that the moment a customer contacts a builder is the instant superior customer service should begin. Customer service should then follow your new home buyer through the financing, construction, closing, customer orientation, move-in, and even for years thereafter. At each critical stage, your customers will have questions, concerns and needs that you must address. Going over and above the call of duty in providing excellent customer service will pay off for you in countless ways, not the least of which are satisfied customers and a steady stream of referral years later. Your goals for a superior customer service program should be the following:

- Set customer service criteria for all of your team.
- Establish quality standards and communicate to your team.
- Know your obligations for providing warranty service and fulfill those obligations.
- Develop repeat and referral sales from satisfied customers.

To learn more about how to provide excellent customer service, take the NAHB course, "Customer Service," which is available in-class or online. The course provides credits toward these NAHB designations: CGA, CGB, CGR, and Master CSP. See nahb.org/education.

Provide a Homeowner's Manual

When construction begins on a customer's home, you should present your customer with an owner's manual. It can be as simple as a series of three-hole-punched handouts secured in a binder or as elaborate as an illustrated, softcover book. You should have your attorney review your home owner's manual for potential liability, especially the warranty section.

Tech Tip

To keep negative comments off of social media, particularly business review sites such as Yelp!, send a postcard with the following verbiage to your customers as they are going through the building process, and again at move-in. Or you can incorporate into one of your existing customer surveys.

> "We know you have choices when it comes to expressing your opinions today. If you have a problem or concern, please share it with us first before you post your issue online, and give us a chance to fix it."

Follow up with a space to fill in with detailed comments and the customers' contact information.

Read More About it

Some good builder resources for home owner manuals include:

- *Home Maintenance Made Easy: What to Do, When to Do It, and When to Call for Help,* by National Association of Home Builders
- *Homeowner Manual: A Template for Home Builders*, by Carol Smith
- *Your New Green Home and How to Take Care of it*—Homeowner Manual Education Template by NAHB Research Department
- *Your New Home & How to Take Care of It.* This is a booklet, available in packs of ten, for builders to handout to new home owners at closing.

All of these items available through NAHB's BuilderBooks at BuilderBooks.com.

See Checklist 13.1, Your Homeowner's Manual, to find out what components your Manual should include.

Checklist 13.1 Your Homeowner's Manual

___ **Introduction.** In this section you should thank your customers for purchasing one of your homes and give them your pledge to provide an outstanding product and superior service throughout the construction phases and beyond. Include names of important contacts such as your sales representative, title company, lender, and design center contacts.

___ **Construction Phases.** This section provides the details about building a customer's new home and highlights the stages at which the customer must make certain decisions, for example, the stage for finalizing selections for flooring, cabinets, landscaping, and other options. It should also address safety concerns, such as the requirement that customers must:

- Have a representative of your team present when walking their homesite in progress
- Take special care when children are present
- Wear hard hats while on-site. (You should provide hard hats for customer use.)

This section might also mention at what stage building officials will make inspections and when the builder will need to obtain various permits.

___ **Change Orders.** This section explains your policy for accepting changes in selections or design along the way. It establishes the deadline for making such changes, shows a sample of your change order form, and states that the customer is likely to incur additional charges for change orders.

___ **Utilities.** Remind the new home owners that they need to contact the local utility and phone companies about establishing service in their names and should do so well in advance of their move-in date. Provide a list of applicable utilities and their phone numbers.

___ **Warranties.** State what your general service warranty will cover, such as materials and workmanship, how long it will be in effect, and what the warranty will not cover, such as appliances and/or surface cosmetic items. It should also outline the procedure for reporting a warranty item. Also note any extended home owner's warranty programs in which you participate to cover major repairs, such as structural problems. Explain that various product and appliance manufacturers represented in the home offer separate warranties as well.

___ **Emergencies.** Outline the procedure for obtaining service during electrical and plumbing emergencies, natural gas leaks, or interruptions in heating or cooling service. Provide a list of utilities and trade contractors and their phone numbers for each of those situations.

___ **Service Requests.** Explain the procedure for requesting a repair or correction. Include a form to fill out, and provide the name and phone number of the person in charge of processing such requests. Offer your pledge to fill requests in as expedient a manner as possible, for instance, within 7 days for non-emergencies.

___ **Maintenance.** Outline various appliances, finishes, and systems that require special or general routine maintenance, such as the heating, air-conditioning, and ventilation system (HVAC), water softeners, solar systems, hardwood floors, gutters, and downspouts. Identify the tasks that are the homeowner's responsibility; for example, check and replace air filters monthly on HVAC, regularly test smoke detectors and electrical receptacles with ground-fault circuit interrupters (GFCI); have the chimney on a wood-burning fireplace professionally cleaned annually; and recaulk around sinks and bathtubs as needed.

Manufacturers' Warranties and Instructions

Compile manufacturers' warranties and care and maintenance information on all appliances and systems in the home. Present these in an attractive container, envelope, or binder and give it to your customer at the orientation. Include surplus decorator items and/or samples for future touch-ups and reordering. Below are some suggestions:

- Hardwood, stone, tile or other flooring
- Carpet
- Stone, tile or marble tub surround
- Countertop material
- Paint, interior and exterior wall and trim
- Roofing tile or shingles
- Siding material
- Tile, brick, or stone fireplace surrounds

Builder's Warranties and Information

Provide the customer with the additional information listed below. (You may have presented some of these items to your buyers at the start of construction with the homeowner's manual. You can remind them of this at the orientation.)

- Builder's warranty
- Information on extended or structural homeowner's warranty programs
- Manufacturers' warranties and care information for the materials noted above
- Contact information for customer service calls
- Names, phone numbers, and addresses of utilities, cable television providers, local telephone company, post office, community newspaper, and recycling centers
- Area map

- Chamber of commerce information on area amenities, and/or brochures from the closest shopping center, theater, art center, and recreation department
- Emergency phone numbers for fire, police, gas and electric companies, the poison control center, and the closest hospital

Conduct a Professional Customer Orientation

The *home owner orientation*, sometimes referred to as the "walk-through" provides the opportunity for you and your customer to inspect the new home together. The meeting also allows you to demonstrate features, highlight benefits, and record any items (also known as a *punch list*) that need correction.

Framing Tour

Some builders, particularly custom builders, may want to conduct a framing tour well before the home owner orientation, and that is when the home is framed out. Let your customer know in advance that during or after framing is not the time to make major changes to the floor plan. Rather, it is an opportunity to demonstrate energy-efficient features (for example, advanced framing techniques and open-web trusses) and their benefits, while confirming that certain items meet the specifications in the sales contract. See Checklist 13.2 to see what should be covered in the framing tour.

Checklist 13.2 Framing Tour

During the framing tour you can accomplish the following tasks:

___ Verify that the floor plan is laid out as specified.

___ Confirm that rooms are of the correct dimensions.

___ Verify that spaces for special features are included, such as oversized tubs, art niches, walk-in closets in the secondary bedrooms, covered porches and balconies, and other items that affect future construction of the home.

___ Maintain the customers' excitement about their purchase and reaffirm that they have chosen the right builder.

___ Provide another opportunity for you or the salesperson to bond with the customer.

___ Offer a final opportunity to sell options and upgrades.

___ Provide an opportunity for the customer to tour their home safely.

___ Verify that hard wiring is included, such as for audio, security, and high-speed data systems.

___ Confirm that plumbing, electrical, phone, cable television and computer outlets are in all desired locations.

Completion Tour

After the home is completed, but prior to settlement, you and the new homeowners should meet so you can show them how to operate appliances and equipment, review maintenance and warranty information, and point out the beauty of the home and its meticulous construction. If a real estate agent was involved in the sale, invite him or her to walk the home with you separately, before the home owner orientation.

Allow several hours for the orientation. It provides one of your best opportunities to celebrate the completion of their new home and to demonstrate your commitment to excellence. Train everyone with a role in finishing the home—including construction, sales, landscaping, and cleaning—to go out of their way to make the home spotless, attractive, and move-in ready. The new homeowners have probably made the largest financial investment they have ever or will ever make; they have every right to expect every last detail to be perfect.

The steps in Checklist 13.3, Prepare for Your Home Owner Orientation, will help you finish final tasks and conduct a thorough inspection to ensure your home shines in time for the customer orientation

Checklist 13.3 Prepare for Home Owner Orientation

Finish Final Tasks

Before the customers arrive for their orientation, complete the following:

___ Have the home cleaned, inside and out.

___ Remove window stickers and clean windows inside and out.

___ Remove construction debris, fallen timber, and litter from the grounds.

___ Make sure landscaping is in or at least rake and level the yard.

___ Sweep porches, decks, driveways, sidewalks, garages, and basements.

___ Mop and/or polish floors (in accordance with manufacturer's directions).

___ Vacuum carpets immediately before the orientation.

___ Clean countertops fixtures, and chrome in the kitchen, bathrooms, and laundry room so they sparkle.

Complete Installations

Install and connect the following items:

___ Closet rods and shelves. Remove sawdust.

___ Screens on windows and patio doors.

___ Light fixtures and ceiling fans. (Have all lights and fans on for the orientation.)

___ Power, including electrical, gas, and water. (Adjust the thermostat well ahead of time so the inside temperature is comfortable during the orientation.)

Test Systems, Equipment, and Appliances

Test, and adjust if necessary, the following items before the orientation. After they are all in working order, demonstrate them all to your customers.

___ HVAC systems

___ Appliances

___ Dishwasher

___ Disposal

___ Microwave oven

___ Range, oven, and exhaust vents and hoods, warming drawers

___ Refrigerator and/or freezer, refrigerator drawers (including water and ice dispensers, if they are included)

___ Washing machine and dryer (if included)

___ Electrical breakers and GFCIs

___ Garage door opener and/or keyless entry

___ Heating system

___ Hot tubs, spas, saunas, pools, fountains

___ Irrigation system

___ Outside faucets

___ Plumbing fixtures, such as sink, shower, bathtub faucets, and toilets

___ Security system

___ Shut-off valves to gas and water

___ Smoke detectors, carbon monoxide detectors

___ Sound systems (play soft non-distracting music during orientation)

___ Water heater

New Home Owner Orientation Form

You can use Form 13.1 to record your customer's comments and requests for corrections during a customer orientation. After you have corrected the items, get your homeowners to sign the form confirming that the work was done to their satisfaction. Put this form on your letterhead with your company's address, logo, name, phone number and email of customer service contacts, and website address.

Superior Customer Service

Executive-level customer service will elevate your stature in your customers' eyes and help you get referrals, one of the best sources for new home sales. Here are some suggestions for providing customer service that is a step above your competition's:

- Send automatic email blasts updating the customer on the construction of their home and allowing them to participate in brief surveys.
- Send a welcome kit to your new homeowners on or shortly before move-in day. Include helpful items such as change-of-address cards, a tube of caulk, tape measure, paint brushes of various sizes, shelf paper and scissors, liquid soap, toilet paper and paper towels, and nontoxic cleaners for any special surfaces in the home, such as granite or marble.
- Appoint one individual to take charge of handling warranty service requests and provide customers with his or her name and contact information. In a small company, the construction superintendent or project

Form 13.1 Sample New Home Owner Orientation Form

[Builder's Letterhead/logo/website]

Homeowner ____________________

Street address ____________________

City, state, zip ____________________

Phone ____________________

Email address ____________________

Community ____________________

Lot # ____________________

Orientation conducted by ____________________

Date of Orientation ____________________

During an orientation and inspection of our new home, we noted the following items requiring correction or completion:

Item	**Room/Location**	**Date Corrected**

[Builder's Name] pledges to the home owner to make every effort to complete the items listed above within 10 days from the above-listed date. In the event of a back-ordered item or other delay, we will inform the homeowner immediately with a revised date of completion.

Builder or Customer
Service Representative ____________________ Date __________

All of the above-listed items have been addressed or completed to our satisfaction.

Homeowner ____________________ Date __________

Homeowner ____________________ Date __________

manager might fulfill this role. In a larger company, there could be one person or an entire department who answers, assigns workers to, and follows up on service requests. (Always update your customers if this role is reassigned.)

- Offer your new homeowners the courtesy of regrouting and recaulking all ceramic tile, bath, sink, tubs, and faucets as needed during the first year (or longer at your discretion).
- Provide a grace period of at least 30 days after your builder's warranty expires.
- If you are a luxury custom builder, consider offering lifetime warranty service on a limited menu of repairs for as long as the original customer owns the home.
- Fill customer service requests within 7 days or immediately in an emergency. If items are back-ordered or you must delay service for another reason, inform the customer immediately and provide the rescheduled date of service.
- Follow up with customers after all service requests are fulfilled to ensure they are satisfied with the work.
- Sixty days before the warranty expires, send a postcard or email (preferably both) reminding the customers that the warranty period is almost up and ask what items still need to be fixed. Send another at 30 days.
- Send a gift basket to your home owners on their one-year anniversary of moving into their new home. This gift will help them remember to send referrals your way.

Conduct Customer Surveys

Customer surveys are a valuable part of your ongoing marketing research. These can be emailed for convenience, including a form that is fillable online and can be returned by clicking "send." Or if your technology does not allow for that, at best, send as an attachment with the option to print and mail, fax, or scan and email back.

Include an incentive for responding or promise a gift upon return of the survey, such as a gift card to a popular local restaurant or attraction. On the follow-up surveys, ask for names of others who might be interested in one of your homes. Offer a referral incentive, such as a free flatscreen TV, iPad, or water filtration system to be installed when the person referred buys a home from you. Form 13.2 is a sample customer service survey. You can adapt it for each subsequent survey.

End-of-Warranty

When your buyers are nearing the end of their warranty period, provide them with a notice at 30 days and at 60 days out, and give them this last opportunity to file any customer service requests. You can use Form 13.3 for this purpose.

Now you should understand the components of a strong customer service program. Turn to the next chapter to find out how you can utilize technology in all areas of your sales and marketing program.

Form 13.2 Sample Customer Service Survey

[Builder's Letterhead]

Please take a few moments to fill out the survey below. Your answers will help us to provide future customers with the highest quality homes possible and to continue to offer you the highest form of customer service. When complete, please return by email, fax or mail.

To thank you for your input, when we receive your completed survey form, we'll send you a gift certificate to ______________________________ Restaurant.

1. What have you enjoyed most about your new home? ______________________________

2. What, if anything, would you like to change about your new home? ______________________________

____

3. Overall, how satisfied have you been with your new home? Circle the number that best represents your satisfaction level, with 5 being the highest.

   5 4 3 2 1

4. Rate the salesperson for each statement using the scale from 1 to 5 shown below:

	Untrue	**Somewhat Untrue**		**Somewhat True**	**True**
A. I felt at ease with the salesperson at all times.	1	2	3	4	5
B. The salesperson was willing to help solve any problems that arose.	1	2	3	4	5
C. The salesperson followed up promptly with answers to questions.	1	2	3	4	5
D. The salesperson was open and helpful during change order requests.	1	2	3	4	5
E. The salesperson was accessible during:					
(1) financing approval process	1	2	3	4	5
(2) construction phase	1	2	3	4	5
(3) decorator selection phase	1	2	3	4	5
(4) customer orientation (walk-through)	1	2	3	4	5

Are there any ways in which your salesperson could improve? ______________________________

____

____

____

Form 13.2 Sample Customer Service Survey (*continued*)

5. Overall, how would you rate the performance of our customer service and warranty program?

   5 4 3 2 1

6. Were any repairs or corrections handled courteously and in a timely manner? ○ Yes ○ No

   If not, please explain: ______________________

   ____

7. Have the repairs or corrections been completed and fixed to your satisfaction? ○ Yes ○ No

   If not, please explain: ______________________

   ____

8. Would you purchase one of our homes again? ○ Yes ○ No

   Why or why not? ______________________

   ____

9. Have you ever referred others to us to buy? ○ Yes ○ No

   If so, did the person purchase a home? ○ Yes ○ No

10. Would you feel comfortable referring others to us to buy ○ Yes ○ No

    Why or why not? ______________________

    ____

11. Would you kindly provide names of three friends who might enjoy living in one or our homes? If one of your referrals buys a home from us and meets the criteria of our customer referral program, we will give you a complimentary ______________________

    A. Name(s) ______________________

       Address ______________ City, state, zip ______________

       Phone number ______________ Email address ______________

    B. Name(s) ______________________

       Address ______________ City, state, zip ______________

       Phone number ______________ Email address ______________

    C. Name(s) ______________________

       Address ______________ City, state, zip ______________

       Phone number ______________ Email address ______________

Thank you for your responses. We value your input.

Form 13.3 Sample End-of-Warranty Service Request

[Builder's Letterhead/Logo/Website address]

Date ______________________________________

Dear ______________________________________

Congratulations! ______________________________, 20xx will mark the one-year anniversary of the purchase of your new ______________________ home. We know your time is valuable, and therefore you may not have had time to place some last-minute service requests you have been meaning to address.

We are inviting you to do that now. Just fill out the information below and return it by email, fax or mail one of the contacts above. If you have any questions, give us a call at (______) ______________ and ask for ______________________________ your Customer Service Representative. We will call you to discuss your request immediately.

Sincerely,

____

President/Owner/CEO/Director of Marketing, etc.

____

Company Name

Item	**Room/Location**

UTILIZE TECHNOLOGY 14

Technology has changed so rapidly that when the first edition of this book was printed, it didn't even include a chapter on technology. At that time, consumers were just beginning to test the waters of web surfing and only the largest or most technologically-savvy builders had an Internet marketing program. Today, a website is a must and an Internet marketing strategy is becoming more important with every passing month for any builder who wants to capture today's customers.

Many buyers today prefer to take control of their home buying experience by tapping the power of the technology in their own home office or family room rather than confronting a salesperson face to face. What's more, prospects who are referred to sales office via the Internet are already prequalified buyers in that they have already chosen the type of location and home they are seeking. This means that there is a higher probability of converting them to buyers. Clearly a builder today must have an effective Internet marketing strategy.

Internet Marketing Begins With an Effective Website

For many potential customers, their first impression of your company is your website. The design of your website is just as crucial to bringing customers to your sales office as the design of the homes you build. An effective website draws them in and entices them to want more information. Some of that will be on your site. But to get more specific information, your site should invite them to contact you whether via email, phone, or visiting your sales office. Chapter 5, Create Your Company Image, discusses what features an effective website should include.

Additionally, having an effective website may allow you to reallocate some of your marketing budget from print and other media toward electronic media. For example, on-site sales agents can use the website in the sales office to navigate through a virtual tour for plans that may not be modeled, and then print an electronic brochure, photos, or floor plans and present them to the customer. Virtual

Figure 14.1 Internet Marketing Strategies

This illustration shows how a builder can integrate all the elements of an online marketing strategy. *Used with permission from David Miles, MIRM, Brand Strategist, Miles BrandDNA.*

models allow the salesperson to demonstrate how a floor plan could look with optional bonus rooms or how kitchen cabinets look with different finishes. Changes to prices and options, as well as other updates, are easier to make on a website and saves the expense of printing new hard copies.

Cater to Your Tech-Savvy Consumers

In addition to reaching your prospects through electronic media, you need to consider another important component of technology. Because these tech-savvy consumers have reached you in the first place, they will likely demand high-tech features in their new homes. The following are suggestions for how to reach the tech-savvy home buying customer:

- Offer separate home offices that don't feel enclosed. Consider offering separate, private entrances or French doors that open onto a side yard or inner courtyard.
- Include as standard computer niches or kitchen "command centers."
- Offer structured wiring systems for home automation
- Offer as options high-tech security systems, video monitoring, surround sound, SMART™ house systems, and home theaters.
- Educate your on-site salespeople and design center personnel about the benefits and features of the hightech components you offer and ensure your on-site salespeople are pointing out these features during model demonstrations.
- Provide technological partners' collateral material in the model home or sales office.

Figure 14.2 Technology Niche

A technology niche in this condo at City Centre in Kitchener, Ontario was carved out of this hallway providing a place for plug-ins, docking station, and email checks. Builder is Andrin Limited and merchandising is by Possibilities for Design.

Used with permission from Doris Pearlman, MIRM, Possibilities for Design.

High-Tech Resources

The following are resources for learning more about increasing the power of technology in your overall marketing plan.

- These two books, available from Builderbooks.com:
  - *Internet Marketing: The Key to Increased Home Sales* by Mitch Levinson
  - *Social Media 3.0: It's Easier Than You Think* by Carol L. Morgan
- *Constructech,* a magazine published six times a year that focuses on builder and technology. See constructech.com
- The NAHB Directory of Professionals with Home Building Designations. See http://tinyurl.com/knb9edq.
- *Sales + Marketing Ideas* is published by NAHB's NSMC six times a year and sent digitally to NSMC members. The magazine covers Internet marketing, social media, and the latest apps for builder salespeople in its regular column, "Tech Zone," as well as in feature articles.

By now you should be well on your way to developing a well-conceived and effective sales and marketing program. With the checklists, forms, examples and resources in this book, you have many critical tools to assist you in becoming a customer-oriented, marketing-savvy, profit-driven home builder.

Good luck!

GLOSSARY

advertorial. Favorable stories in the media written by advertisers and packaged to look like news.

bleed. A printing technique where the graphic or photograph continues off the edge of the page, as opposed to staying within the perimeter of the paper.

blog. Short for web log, this is a regular, online column written by an individual or company representative to provide content of interest to a specific audience. This is one method of increasing your online presence. Regularly contributing to a blog will provide content for search engines, which will drive greater traffic to your website.

deboss. A printing technique where the text and/or graphics is depressed into the paper.

die-cut. A process where a design is cut into the paper in a particular pattern or shape. Promotional pieces with a die-cut can be more expensive to produce because of their one-of-a-kind nature.

dingbat. Symbols or characters in typesetting, similar to emoticons, that add levity, project mood, or inject fun into a collateral piece. There are several dingbat fonts that use symbols or shapes in lieu of letters and numbers.

email blasts. Email messages sent regularly to your prospect list, also known as "eBlasts" or "drip campaign."

emboss. A printing technique where some of the text and/or graphics is raised in relief on the paper.

feng shui. A philosophy dating back to an ancient Chinese belief that the design and location of one's home and the placement of objects and furnishings within the home will increase or decrease positive spiritual energy. Feng shui is an important consideration in multicultural marketing.

focus group. A market research method to acquire qualitative data using a discussion group of carefully chosen participants gathered together to ask for their responses and opinions.

foil-stamp. A printing process where the graphic or text is coated with a glossy paper.

frequency. The number of times an advertising message is repeated.

home owner orientation. Sometimes referred to as the "walk-through," this is the opportunity for builder and customer to inspect the new home, demonstrate features, highlight benefits, and record any items that need correction.

lead. The introductory paragraph to a press release or advertorial, providing a "hook" to the reader and answering six questions: "Who?" "What?" "When?" "Where?" "Why?" and "How?"

market research. Gathering information about various aspects of the target area where a company does business, including potential customers, their lifestyles and preferences, as well as the competition.

masthead. The part of a newsletter or magazine that credits the editor(s) and contributor(s) and provides contact information. Sometimes the name of the publication and its logo is also referred to as a masthead.

metropolitan statistical area. For the purposes of economic data, an area typically around a large city, that is economically tied to that city that includes surrounding areas as well. (E.g., Chicago) " geographic entity delineated by the Office of Management and Budget for use by federal statistical agencies. Metropolitan statistical areas consist of the county or counties (or equivalent entities) associated with at least one urbanized area of at least 50,000 population, plus adjacent counties having a high degree of social and economic integration with the core as measured through commuting ties."

merchandising. Equipping a builder's model or inventory home with carefully selected items to appeal to the intended target market, enhance the architectural features of the home, and downplay or disguise problem areas.

mystery shopping. A sales training method whereby a professional evaluator posing as a customer calls on your on-site sales professionals to evaluate their sales performances and improve their skills.

promotions. Limited campaigns tied to a one-time, on-site event or a special program, prices, financing, or other incentive.

psychographics. Research data based on lifestyle factors, rather than demographic factors, for example, identifying people who describe themselves as "avid golfers," instead of identifying themselves as "two-income households with children."

punch list. A record of any and all items that are noted as needing attention or correction during the home owner orientation.

qualitative data. A type of data gathered when a researcher is looking for information beyond numbers and percentages (quantitative data), but wants to learn about respondents' opinions, practices, and answers to open-ended questions.

reach. The number of prospects who will receive a single advertising message.

return on investment (ROI). A measure of profitability versus the amount of time and money invested.

search engine optimization (SEO). Using various tools and techniques to ensure a website appears high in search results for a variety of keywords and phrases.

serifs. Tiny finishing edges on certain fonts, such as Times New Roman. Serif fonts are considered easier to read in print than sans serif fonts. An example of a sans serif font would be Arial.

troikas. A term coined by the late Briggs Napier, MIRM, it refers to small informal discussion groups of about three participants. Putting the results of three or four troikas together over a period of time will provide information similar to that of a larger focus group and can be more cost-effective than conducting one or two a large focus groups.

trompe l'oiel. Literally translating to "fool the eye" in French, this refers to a wall painting technique or mural that resembles a window, doorway, view or other scene, and can be used in vignetting a home.

varnish. A paper-coating procedure that results in a glossy finish. Can be done for all or a portion of a promotional piece.

vignetting. To demonstrate the use of rooms in an inventory or model home that is not fully furnished with small decorator touches and accents that help the home appeal to the intended target market.

Web log. *See* blog.

xeriscaping. A method of landscaping that uses little or no water for desert or drought-prone climates.

NOTES

Introduction

1. "The Digital House Hunt: Consumer and Market Trends in Real Estate, A Joint Study from The National Association of REALTORS® and Google," Realtor.org, Jan. 16, 2013, www.realtor.org/reports/digital-house-hunt.

2. Sabrina Tavernise, "Married Couples Are No Longer a Majority, Census Finds," *New York Times,* May 26, 2011, www.nytimes.com/2011/05/26/us/26marry.html?_r=0.

Chapter 3

3. Jessica Lautz, "Median Age of Home Buyers: 2001–2010," *Economists' Outlook* (blog), March 7, 2011, http://economistsoutlook.blogs.realtor.org/2011/03/07/median-age-of-home-buyers-2001-2010/.

4. Robert Krueger, "Where Americans Want to Live: New ULI Report, *America In 2013*, Explores Housing, Transportation, Community Preferences Survey Suggests Strong Demand for Compact Development," May 15, 2013, www.uli.org/press-release/america2013/.

5. The Futures Company, *Global Monitor Survey Report,* http://thefuturescompany.com/free-thinking/unmasking-millennials/.

6. "Baby Boomers Control 70% of U.S. Disposable Income," MarketingCharts.com, Aug. 7, 2012, www.marketingcharts.com/television/baby-boomers-control-70-of-us-disposable-income-22891/.

7. "Baby Boomers Control 70% of U.S. Disposable Income," MarketingCharts.com, Aug. 7, 2012, www.marketingcharts.com/television/baby-boomers-control-70-of-us-disposable-income-22891/.

8. AARP Florida, "The Sandwich Generation: You Are Not Alone," AARP.org, June 28, 2012, www.aarp.org/home-family/caregiving/info-06-2012/sandwich-generation-fl1845.html.

9. U.S. Census Bureau, The 2012 Statistical Abstract, www.census.gov/compendia/statab/cats/international_statistics/population_households.html.

10. Laryssa Mykyta, "Economic Downturns and the Failure to Launch: The Living Arrangements of Young Adults in the U.S. 1995–2011," U.S. Census Bureau SEHSD Working Paper 2012-24, www.census.gov/hhes/www/poverty/publications/WP2012-24.pdf.

Chapter 5

11. "The Digital House Hunt: Consumer and Market Trends in Real Estate, A Joint Study from The National Association of REALTORS® and Google," Realtor.org, Jan. 16, 2013, www.realtor.org/reports/digital-house-hunt.

Chapter 6

12. "Baby Boomers Control 70% of U.S. Disposable Income," MarketingCharts.com, Aug. 7, 2012, www.marketingcharts.com/television/baby-boomers-control-70-of-us-disposable-income-22891/.

Chapter 8

13. Housing and Urban Development, Fair Housing Advertising, Part 109, www.hud.gov/offices/fheo/library/part109.pdf.

Chapter 12

14. U.S. Census Bureau, "Language Other than English Spoken at Home," State and County Quick Facts, 2012, http://quickfacts.census.gov/qfd/meta/long_POP815212.htm.
15. Steven A. Camarota, "Immigrants in the United States, 2010: A Profile of America's Foreign-Born Population," CIS.org, August 2012, www.cis.org/2012-profile-of-americas-foreign-born-population.
16. Steven A. Camarota, "Immigrants in the United States, 2010: A Profile of America's Foreign-Born Population," CIS.org, August 2012, www.cis.org/2012-profile-of-americas-foreign-born-population.
17. Steven A. Camarota, "Immigrants in the United States, 2010: A Profile of America's Foreign-Born Population," CIS.org, August 2012, www.cis.org/2012-profile-of-americas-foreign-born-population.
18. U.S. Census Bureau, "Asians Fastest-Growing Race or Ethnic Group in 2012," news release, June 13, 2013, www.census.gov/newsroom/releases/archives/population/cb13-112.html.
19. Jeff Collins, "Homebuilders on board with feng shui," *The Orange County Register,* May 21, 2012, www.ocregister.com/articles/feng-354925-shui-home.html.
20. U.S. Census Bureau, "Hispanic Origin," State and County Quick Facts, 2012, http://quickfacts.census.gov/qfd/meta/long_RHI725211.htm.
21. U.S. Census Bureau, "U.S. Census Bureau Projections Show a Slower Growing, Older, More Diverse Nation a Half Century from Now," news release, Dec. 12, 2012, https://www.census.gov/newsroom/releases/archives/population/cb12-243.html.
22. "Money: What do they Earn and Spend?," ReachingBlackConsumers.com, Accessed December 11, 2013, www.reachingblackconsumers.com/money.
23. "The People: Who Are They?," ReachingBlackConsumers.com, Accessed December 11, 2013, www.reachingblackconsumers.com/the-people.
24. "2010 Census Shows Interracial and Interethnic Married Couples Grew by 28 Percent over Decade," April 25, 2012, Census.gov, www.census.gov/newsroom/releases/archives/2010_census/cb12-68.html
25. Leanne Italie, "Cheerios Exec On Ad Featuring Mixed Race Couple: 'We Were Reflecting An American Family,'" Huffington Post.com, June 20 2013, www.huffingtonpost.com/2013/06/05/cheerios-ad-mixed-race-couple_n_3390520.html.
26. Todd Wasserman, "How to Reach African-American Consumers" *Adweek,* January 14, 2010, www.adweek.com/news/advertising-branding/how-reach-african-american-consumers-106945?page=2.

INDEX